Mathematics – The Music of Reason

Springer
Berlin
Heidelberg
New York
Barcelona
Budapest
Hong Kong
London
Milan
Paris
Singapore
Tokyo

Jean Dieudonné

Mathematics –
The Music of Reason

Translated by H.G. and J.C. Dales
With 41 Figures

Springer

Author:
Jean Dieudonné †

Translator:
H.G. and J.C. Dales
Department of Pure Mathematics
University of Leeds
Leeds, LS2 9JT
England

Title of the original French edition:
Pour l'honneur de l'esprit humain © 1987 by Hachette
ISBN 2-01-011950-9

ISBN 978-3-642-08098-2

Corrected Second Printing 1998

Springer-Verlag Berlin Heidelberg New York

Mathematics Subject Classification (1991): 00-XX, 00A05, 00A06, 00A30, 00A99, 01-01

Library of Congress Cataloging-in-Publication Data
Dieudonné, Jean Alexandre, 1906- [Pour l'honneur de l'esprit humain. English] Mathematics – The music of reason: with 41 figures / Jean Dieudonné. p. cm. Includes bibliographical references and index. 1. Mathematics. I. Title.
QA36.D5413 1992 510-dc20 92-27390

© Springer-Verlag Berlin Heidelberg 2010
Printed in Germany

«... M. Fourier avait l'opinion que le but principal des mathématiques était l'utilité publique et l'explication des phénomènes naturels; mais un philosophe comme lui aurait dû savoir que le but unique de la science, c'est l'honneur de l'esprit humain, et que sous ce titre, une question de nombres vaut autant qu'une question du système du monde.»

C.G.J.JACOBI.
Lettre (en français) à Legendre, 2 juillet 1830.
Gesammelte Werke, Vol. I, Berlin (Reimer), 1881, p. 454.

"... Monsieur Fourier was of the opinion that the principal aim of Mathematics is to serve mankind and to explain natural phenomena; but a philosopher such as he ought to have known that the sole aim of science is the fulfillment of the human spirit, and that, accordingly, a question about numbers has as much significance as a question about the workings of the world."

C.G.J.JACOBI.
Letter (in French) to Legendre, July 2, 1830.
Gesammelte Werke, Vol. I, Berlin (Reimer), 1881, p. 454.

"May not Music be described as the Mathematic of Sense, Mathematics as Music of the Reason? The soul of each the same?"

J.J.SYLVESTER.
Algebraical Researches, containing a Disquisition on Newton's Rule for the Discovery of Imaginary Roots.
In: Philosophical Transactions of the Royal Society, Vol. 154, 1864, pp. 579–666.

To Odette and Françoise

Contents

Introduction

This work is designed expressly for readers who for various reasons are interested in science, but *are not professional mathematicians*. Experience shows
that it is almost always the case that while such people enjoy reading or listening to expositions on the natural sciences, and feel that they are gaining
from them knowledge which enriches their view of the world, they find an
article on current mathematics to be written in an incomprehensible jargon,
and dealing with concepts too abstract to be of the slightest interest.

The object of this book is to try to explain the reasons for this lack of understanding, and perhaps to do away with it. Since it is not possible to discuss
mathematics without taking for granted a certain minimum of mathematical
knowledge, I have to assume that my readers will have followed a course in
the subject to the level of the *scientific baccalaureate*[1]. I have inserted some
additional material, in the shape of appendices to the chapters, which is intended for those who have pursued their scientific studies a little further (at
least up to the end of the second year at a university); but it is not necessary
to read *any* of this additional matter in order to understand the main text,
and so it can be omitted without detriment. All that is necessary is a slight
effort of sustained concentration in following a chain of reasoning, each link
in which, taken by itself, is entirely elementary.

What I intend to show is that the very *nature* of this 'minimum luggage'
is at the bottom of the difficulties experienced in understanding mathematics.
*Nothing which is taught at a secondary school was discovered later than the
year 1800*[2]. The case is not very different when we come to the mathematical
knowledge deemed necessary for all branches of natural science other than
physics. Even among physicists, apart from those who work in quantum theory
or relativity, I believe that those who do experimental work use hardly any
more mathematics than was known to Maxwell in 1860.

Everybody knows that the natural sciences have advanced continuously
in the most extraordinary way since the beginning of the nineteenth century.

---

[1] This is an examination in France taken when students leave school, usually at age
19, that determines whether they can enter university. It is approximately equivalent to "A – levels" in a group of scientific subjects which includes mathematics,
in England and Wales, and its standard is roughly that of the end of an Honours
course in freshman calculus, taken in the first year of university in the U.S.A.
(Translator).

[2] But not *everything* which was known before this date is taught there – far from it!

The 'educated' public has been able to follow this dizzying ascent, a few paces behind, certainly, but without losing its footing, thanks to clever simplifications which conserve what is essential in the new ideas. Mathematics has advanced in just the same way, but, apart from mathematicians, hardly anybody is aware of the fact. This is because, on the one hand, as was said above, most of the sciences, even today, call for scarcely anything beyond 'classical' mathematics[3]; on the other hand, there has occurred in the evolution of mathematics a genuine mutation: new 'mathematical objects' have been created, quite different from the 'classical' objects, numbers and 'figures'. The far more severely abstract nature of these objects (since they are not supported in any way by visual 'pictures') has deterred those who did not see any use in them.

I would like to convince the well-disposed reader that this abstraction which has accrued in no way sprang from a perverse desire among mathematicians to isolate themselves from the scientific community by the use of a hermetic language. Their task was to find solutions for problems *handed down by the 'classical' age*, or arising directly from new discoveries in physics. They found that this was possible, but only through the *creation* of new objects and new methods, whose abstract character was *indispensable* to their success.

I shall therefore try to show that mathematicians from 1800 to 1930 were *constrained* by the essential nature (often unrecognized until then) of *classical* objects and relations to forge new 'abstract' tools which enabled them and their present-day successors to solve problems which before seemed unassailable. Contemporary mathematicians have been able to make imaginative use of these tools and to extend their range, but, contrary to what is sometimes thought, they did not create them.

I do not think it is possible to understand contemporary mathematics without having at least some idea of the main features of its past. The first two chapters, outlining the position of mathematics and mathematicians in today's world, as well as the connections between mathematics and the other sciences, will be followed by a brief survey in Chapter III of classical mathematics from Euclid to Gauss. This chapter will lay stress on the great original idea of the Greeks: that is, to ascribe to mathematical objects, in a way which has proved ineffaceable, the character of 'objects of thought'. Chapter IV gives a few examples, chosen from among thousands, of problems which occurred in classical mathematics and which gave rise to important research during the nineteenth century.

Chapter V is the central point of my argument. Again, by means of a few examples accessible to the reader, we see how, in the nineteenth century, the analysis of problems inherited from the previous century led gradually to the discovery of their true nature, thus paving the way to their partial or complete solution. But the price to be paid is the necessity of abandoning the semi-'concrete' character of classical mathematical objects; it has to be understood that what is essential about these objects is not the particular features which they seem to have but the *relations* between them. These relations are often

---

[3] I give this name to all results known before 1800.

the same for objects which appear to be very different, and therefore they must be expressed in ways which do not take these appearances into account: for example, if we wish to specify a relation which can be defined either between numbers or between functions, it can only be done by introducing objects which are neither numbers nor functions, but which can be *specialized* at will as either numbers or functions, or indeed other kinds of mathematical objects. It is these 'abstract' objects which are studied in what have come to be called mathematical *structures*, the simplest of which are described in Chapter V.

Finally, Chapter VI tells of the difficulties which arose in the nineteenth century in finding a way to formulate definitions and theorems with such complete precision as would leave no room for uncertainty. To this end, Euclid's *Elements* was used as a model in all branches of mathematics, once its inaccuracies had been corrected and once the purity of the Greeks' original conceptions had been rediscovered; something which had been too long neglected in the euphoria induced by the incredible successes attending the application of mathematics to the experimental sciences[4].

Lovers of sensational books will be disappointed in this one; they will find here no world-shattering theses, bristling with ingenious paradoxes. I have limited myself to stating *facts* concerning mathematics, not opinions, and I have taken care not to engage in polemics. I have equally carefully refrained from attempting any predictions as to the future evolution of the concepts I am dealing with.

Annotated Bibliography

[1] J. Dieudonné, and 10 collaborators: Abrégé d'Histoire des Mathématiques 1700–1900, new edition, Paris (Hermann), 1986.

[2] G. Choquet: La Naissance de la Théorie des Capacités: Réflexion sur une Expérience Personnelle, "La Vie des Sciences", C.R. Acad. Sci., Série Générale, 3 (1986), pp. 385-397.

[3] P. Dedron and J. Itard: Mathematics and Mathematicians, 2 vol., trans. from the French by J.V. Field, Milton Keynes (Open University Press), 1974.

[4] Diophantus of Alexandria: Les Six Livres d'Arithmétique et le Livre des Nombres Polygones, trans. P. Ver Ecke, Bruges (Desclées de Brouwer), 1926.

[5] Euclid: The Thirteen Books of Euclid's Elements, 3 vol., trans. T. Heath, New York (Dover), 1956.

[6] L. Euler: Correspondance avec C. Clairaut, J. d'Alembert et J.-L. Lagrange, Opera Omnia, Ser. Quarta A. Basle (Birkhäuser), 1980.

[7] G.H. Hardy: A Mathematician's Apology, 1st edition, Cambridge (Cambridge University Press), 1940.

---

[4] Even physicists are amazed by the 'unreasonable efficaciousness of mathematics' for constructing models adapted to the world of experience, as E. Wigner says.

[8] S. Lang: The Beauty of Doing Mathematics, New York, Berlin, Heidelberg, Tokyo (Springer), 1985.

[9] H. Poincaré: a) La Science et l'Hypothèse, Paris (Flammarion), 1902; b) La Valeur de la Science, Paris (Flammarion), 1905; c) Science et Méthode, Paris (Flammarion), 1908; d) Dernières Pensées, Paris (Flammarion), 1913.[5]

[10] A. Weil: Oeuvres Scientifiques, 3 vol. New York, Berlin, Heidelberg, Tokyo (Springer), 1979.

Since I am not writing for scholars, I have not thought fit to encumber my text with a multitude of 'references', and I have limited the bibliography to a few works in keeping with the spirit of this book. I hope that readers will take me at my word when I assure them that I could cite a text for every one of my assertions relating to the history of mathematics.

It would be impossible to go too far in encouraging readers to become acquainted with more details of this history. I would like to draw attention to book [3], which is rendered attractive by numerous illustrations and requires no knowledge beyond the baccalaureate. At a more advanced level (for those with two or three years of university study), I may be permitted to point out item [1] for the history of the subject since 1700.

It is always instructive to read Diophantus [4] and Euclid [5]; the English translation of Euclid by T. Heath [5] has copious notes and explanations of a mathematical or philological nature.

There are not many mathematicians who have published reflections on their subject based on their personal experience. The huge correspondence of Euler [6] enables us to get to know better his life and his methods of work, and it is always interesting to read the many articles by Poincaré on the subject, collected into four volumes [9]. In a generation closer to ours, G.H. Hardy, at the end of his life, wrote a very appealing *Apology* [7], in which the man is constantly showing beneath the mathematician.

One of the greatest mathematicians of my generation, A. Weil, had the happy idea of accompanying the collected edition of his *Works* [10] with quite extensive commentaries on each of his books or articles: this amounts to an exciting 'working diary', describing almost from day to day the evolution of the thought of a creative mathematician. Even if an unusually extensive mathematical culture is required to appreciate it fully, no-one can fail to to profit from these reminiscences, free as they are from either bragging or false modesty.

Altogether similar and just as enriching is the description recently produced by G. Choquet [2] of the way in which he was led to one of his finest discoveries, that of generalized capacities.

---

[5] Of these works by Poincaré the following are available in English: Science and Hypothesis [trans. W.J.G.], with a preface by J. Larmor, London (Walter Scott Publishing Company), 1905; Science and Method, trans. Francis Maitland, with a preface by the Hon. Bertrand Russell, London (Thomas Nelson and Sons) 1914 (Translator)

Finally, I am eager to point out a recent heroic endeavour by the Franco-American S. Lang to bring to the unaccustomed public of the Palais de la Découverte some understanding of what is meant by 'the beauty of mathematics', in three conferences followed by discussions [8]. The willingness of the speaker – even though he is a 'pure' mathematician – to put himself at the disposal of his listeners, and his imperturbability in the face of their unpredictable reactions, make this a sketch as lively and entertaining as it is instructive.

I Mathematics and Mathematicians

1. The Concept of Mathematics

The position of mathematics among human activities is paradoxical. In our day almost everybody in the "developed" countries knows that it is an important discipline, needed for most branches of science and technology, and that an ever greater number of professions cannot be exercised without some knowledge of it. Nevertheless, if you ask "What is mathematics?" or "What does a mathematician do?", you will very rarely get anything but a preposterous answer, unless your interlocutor has had a mathematical training at least up to the end of the first two years at a university. Even men who are eminent in other sciences often have only aberrant notions about the work of mathematicians.

Since everybody has made contact with mathematics by means of numerical calculations at primary school, the most widespread idea is that a mathematician is someone who is a virtuoso in these calculations. With the advent of computers and their language, people will now think that that means somebody who is especially good at "programming" them, and who spends all his time doing this. Engineers, always looking for optimal values for the measures of magnitudes which interest them, think of mathematicians as custodians of a fund of formulae, to be supplied to them on demand. But the most fallacious of current conceptions, firmly held by nearly all our contemporaries, inundated as they are by plentiful descriptions in the media of the progress made in all other sciences, is that there is nothing more to be discovered in mathematics, and that the work of mathematicians is limited to passing on the legacy of past centuries.

Must we rest content with this state of affairs, and do no more than remark, with A. Weil ([10], II, p.465) that "mathematics has this peculiarity, that it is not understood by non-mathematicians"? I think that we must at least try to find the reasons for this lack of understanding. There are in fact numerous periodicals devoted to popularizing recent scientific discoveries, addressed to large numbers of readers and at all levels of scientific education. With a few exceptions, to which we shall return (Chapter VI, §5, C), there is nothing of the kind on recent progress in mathematics (which could make people think that none has taken place!); by contrast, there is plenty of literature on astrophysics, geology, chemistry, molecular biology, even atomic or nuclear physics. It seems that it is possible, with the help of diagrams in which details are eliminated and the essential objects clearly displayed, in addition to explanations which simplify radically what specialists have been able to deduce from complex and delicate experiments, to give readers of these publications the illusion that they understand what an atom, or a gene, or a galaxy is, even if the orders of magnitude concerned, ranging from an angström (10^{-10} metres) to a light year ($3 \cdot 10^{15}$ metres), are beyond the grasp of our imaginations.

By way of contrast, let us take one of the most fruitful theories of modern mathematics, the one known as "sheaf cohomology". Thought of in 1946, it is more or less contemporary with the "double helix" of molecular biology, and has led to progress of comparable magnitude. I should, however, be quite unable to explain in what this theory consists to someone who had not followed at least the first two years of a course in mathematics at a university. Even for an able student at this level the explanation would take several hours; while to explain the way in which the theory is used would take a good deal more time still. This is because here we can no longer make use of explanatory diagrams; before we can get to the theory in question we must have absorbed a dozen notions equally abstract: topologies, rings, modules, homomorphisms, etc., none of which can be rendered in a "visual" way.

The same remarks can be applied to almost all ideas which form the basis for the great mathematical theories of our day, a list of which can be found in Chapter V, §5, A. I must therefore resign myself: if I want to discuss what is accessible at baccalaureate level, I can allude only to elementary concepts under three of these headings (out of twenty-one): algebra, number theory and set theory; adding from time to time a remark addressed to readers who have some familiarity with the calculus. Beginning with these notions, I shall try to explain how they have evolved *of necessity*, revealing underlying notions which are much more abstract, and acquiring in the process an incomparably greater efficacy in solving mathematical problems. To go further would mean putting together streams of words in which the non-mathematical reader would very soon lose his footing. The list I have mentioned is there only to give an idea of the *minute* number of theorems which I shall discuss, compared with the immensity of modern mathematical knowledge.

2. A Mathematician's Life

The term "musician", in common speech, can mean a composer, a performer or a teacher of music, or any combination of these activities. In the same way, by "mathematician" we can understand a teacher of mathematics, one who uses mathematics, or a creative mathematician; and the common belief is that this last species is now extinct. Since, however, it is my intention to help my readers to understand the origin and nature of modern mathematics (of which, as we shall see [Chapter V, §5, A], the greater part dates from no earlier than 1840), we shall be dealing almost exclusively with this last category. A mathematician, then, will be defined in what follows as someone *who has published the proof of at least one non-trivial theorem* (the meaning of the expression "trivial" will be precisely defined later [§4]).

It seems to me that it would be not uninteresting for the reader to be taken on a little trip to the unknown country of mathematicians before we try to explain what they do.

There is every reason to think that an aptitude for creating mathematics is independent of race, and it does not seem on the whole to be transmitted by heredity: mathematicians who are the offspring of mathematicians have always been rare exceptions. But in the flowering of a mathematical talent social environment has an important part to play. There is no example, even among the greatest geniuses, of a mathematician discovering for himself, *ab ovo*, the mathematics known to his generation, and the story told of Pascal relating to this is only a legend. The social milieu, then, must be such as to enable the future mathematician to receive at least some elementary instruction exposing him to genuine proofs, awakening his curiosity, and after that enabling him to come into contact with the mathematics of his time: something which depends on his having access to the works where it is written down. Up until the end of the eighteenth century, there was no real organized university education in mathematics, and, from Descartes and Fermat to Gauss and Dirichlet, nearly all the great mathematicians trained themselves without instructors, by reading the works of their illustrious predecessors (still an excellent exercise!). Even today it is likely that many mathematical talents never come to light, for lack of a favourable social environment, and it is not surprising that we do not know of mathematicians in less "developed" societies. Even in countries which have evolved further, the type of elementary instruction can be unpropitious for the flowering of mathematical vocations when it is subject to religious or political constraints, or to exclusively utilitarian concerns, as was the case in the United States until well into the twentieth century. Secondary education there is in effect in the hands of local authorities, who often used to think it was more important for adolescents to learn typing or how to drive a car than Latin or Euclidean geometry. I have known famous American mathematicians who were born in small towns and had never seen a mathematical proof until they were 18 years old; attracted to science, they were oriented at the university towards a career in engineering, and it was only by chance that they were confronted with real mathematical instruction, and thus discovered their true vocations.

In societies where education is available to a sufficiently wide section of society (if necessary with the help of grants), the social origins of mathematicians are very varied. They may come from the nobility, like Fagnano, Riccati and d'Alembert, from the upper middle class, like Descartes, Fermat, Pascal, Kronecker, Jordan, Poincaré and von Neumann, or they may come from very humble backgrounds, like Roberval, Gauss, Elie Cartan and Lebesgue. Most are born into middle-class families, often quite penurious ones.

A mathematical vocation is most often awakened at about fifteen years of age, but it can be held back by an education which does not allow for any concept of proof, as was the case in the United States, as stated above. In any case, contrary to a fairly widespread opinion, the creative period rarely begins before the age of twenty-three to twenty-five; the cases of Pascal, Clairaut, Lagrange, Gauss, Galois, Minkowski, and, in our time, Deligne, who were responsible for notable mathematical discoveries at less than twenty years old, are exceptional. As for mathematical longevity, another current opinion is that, as Hardy said ([7], p.10), "mathematics is a young man's game". It is true that a number of major discoveries have been made by mathematicians who were not much older than thirty. Still, for many of the greatest, such as Poincaré, Hilbert or H. Weyl, the creative period is prolonged, and they remain quite productive up to the age of fifty or fifty-five. Others, such as Killing, Emmy Noether, Hardy (as he himself recognized ([7], p.88)) or more recently Zariski and Chevalley, give of their best when they are about 40. Finally, a few mathematicians favoured by the gods, like Kronecker, Elie Cartan, Siegel, A. Weil, J. Leray, and I. Gelfand, have continued to prove fine theorems after the age of sixty. All the same, just as in many sports one can hardly expect to remain a champion after thirty, so most mathematicians have to resign themselves to seeing their creative imaginations become barren after the age of fifty-five or sixty. For some it is a heartbreaking experience, as Hardy's *Apology* movingly witnesses; others adapt by finding an activity in which they can exercise those faculties which they still possess.

As with many scholars, the life of a mathematician is dominated by an insatiable curiosity, a desire bordering on passion to solve the problems he is studying, which can cut him off completely from the realities around him. The absent-mindedness or eccentricity of famous mathematicians comes simply from this. The fact is that the discovery of a mathematical proof is usually reached only after periods of intense and sustained concentration, sometimes renewed at intervals over months or years before the looked-for answer is found. Gauss himself acknowledged that he spent several years seeking the sign of an algebraic equation, and Poincaré, on being asked how he made his discoveries, answered "by thinking about them often"; he has also described in detail the unfolding of the course of reflections and attempts which led to one of his finest results, the discovery of "Fuchsian" functions [9].

What a mathematician seeks above all, then, is to have at his disposal enough time to devote himself to his work, and that is why, since the nineteenth century, they have preferred careers in teaching in universities or colleges, where the hours given to teaching are relatively few and the holidays are long. Remuneration is of only secondary importance, and we have recently seen, in the United States among other places, mathematicians abandoning lucrative posts in industry in order to return to a university, at the cost of a considerable drop in salary.

Moreover it is only recently that the number of teaching posts in universities has become fairly large. Before 1940 it was very restricted (less than one hundred in France), and up until 1920 mathematicians of the quality of Kummer, Weierstrass, Grassmann, Killing and Montel had to rest content with being teachers in secondary education during all or part of their working lives; a situation which persisted for a long time in small countries with few universities.

Before the nineteenth century, a mathematician's opportunities to do his work were even more precarious, and failing a personal fortune, a Maecaenas or an academy to provide him with a decent living, there was almost no other resource than to be an astronomer or a surveyor (as in the case of Gauss, who devoted a considerable amount of time to these professions). It is this circumstance which no doubt explains the very restricted number of talented mathematicians before 1800.

This same need to devote long hours of reflection to the problems which they are trying to solve almost automatically precludes the possibility of driving in tandem absorbing tasks in other areas of life (such as administration) and serious scientific work. The case of Fourier, Prefect of Isère at the time when he was creating the theory of heat, is probably unique. Where we find mathematicians occupying high positions in administration or in government, they will have essentially given up their research during the time that they are discharging these functions. Besides which, as Hardy commented ironically ([7], p.12), "the later records of mathematicians who have left mathematics are not particularly encouraging".

As regards the personal traits of mathematicians, if we except their absent-mindedness and the lack of "practical know-how" with which many are afflicted, it can be said that there is hardly any difference between them and any other section of the population of comparable status taken at random. The same variety in their characters is to be found, the same virtues and vices. If they have at times a tendency to pride themselves on the possession of a rather uncommon talent, they must remember that many other techniques can be performed by only a small number of people. No doubt there are far fewer tight-rope walkers who can dance on the high wire, or chess-players who can play blind-fold. Nevertheless, as Hardy says ([7], p.16), what a mathematician does "has a certain character of permanence", and you have not lived in vain if you have added a small stone to the pyramid of human knowledge.

In any case, mathematicians must accept that their talent does not confer on them any particular competence outside their own domain, as those who have thought that their prestige among their colleagues would enable them to reform society have learnt to their cost.

The opinions of mathematicians in matters of of religion or politics are moreover very diverse: Cauchy was a bigot, but Hardy was a curious kind of atheist for whom God was a personal enemy; Gauss was very conservative, but Galois was an ardent revolutionary. An extreme and rather distressing case is that of the young German mathematician O. Teichmüller, who perhaps had a genius comparable to that of Galois; but he was a fanatical Nazi, who contributed to the expulsion of his Jewish teachers. He went to the Russian front as an SS officer, and did not return.

Most mathematicians lead the lives of good citizens, caring little about the tumults which convulse the world, desiring neither power nor wealth, and content with modest comfort. Without actively seeking honours, most are not indifferent to them. In contrast to those of many artists, their lives are rarely thrown into chaos by emotional storms; lovers of sensation and romance have little to glean among them, and have to content themselves with more or less romanticized biographies of Galois, of Sonia Kowalewska or of Ramanujan.

3. The Work of Mathematicians and the Mathematical Community

Research in the experimental sciences is done in laboratories, where larger and larger teams are needed to manipulate the instruments and to scrutinize the results. To do research in mathematics nothing is needed except paper and a good library. Team-work, as practised in the experimental sciences is, then, quite unusual in mathematics, most mathematicians finding it difficult to think seriously except in silence and solitude. Collaborative work, while quite common, most often consists in putting together results that each of the collaborators has managed to obtain in isolation, albeit with mutual profit from each other's ideas, enabling them to progress from new points of departure. The best-known example of prolonged collaboration between mathematicians is that of Hardy and Littlewood: one of them lived in Oxford and the other in Cambridge, and they saw each other only occasionally; their work together being done entirely through correspondence.

But if most publications are the work of individuals, those mathematicians are rare who, like Grassmann, Hensel or Elie Cartan, can work productively for a long time in almost complete isolation (due, in these three cases to the novelty of their ideas, which were not understood by their contemporaries). Most become discouraged if they lack the opportunity to communicate quite frequently with their colleagues in the hope that they will understand them, the more so because the abstract nature of their research makes it difficult to exchange ideas with non-mathematicians.

Until the middle of the seventeenth century the only means of communication was private correspondence (sometimes centralized by well-disposed letter-writers, such as Mersenne or Collins), personal visits, and sometimes the printing of a work at the author's expense (unless some Maecaenas consented to pay). The first publications to be regularly issued by academies appeared about 1660, but the numbers were still small, and the few scientific journals which were published before 1820 were devoted to science in general, without specialization. The first specialist mathematical journals to cross national boundaries were the *Journal de Crelle*, founded in 1826, followed closely by the *Journal de Liouville* in 1835.

It was now, however, that linguistic barriers began to arise. In the seventeenth century almost all scientific publications were produced in Latin (Descartes' *Géométrie*, with some of Pascal's writings, being the only noteworthy exceptions). This tradition began to decline in the eighteenth century; French mathematicians now wrote only in their own language, and abroad some works of an instructional nature were written in their authors' languages. After Gauss's time, the use of Latin rapidly disappeared in scientific publications; and this continued to slow down the dissemination of new results, especially in France, where there was hardly any study of living languages until the end of the nineteenth century. A rather ridiculous example occurred when the Paris Academy of Science, in 1880, set up a competition to solve a problem in number theory which had already been solved more than 20 years before by H.J.S. Smith. Apparently Hermite and Jordan did not read English. In our day, as is well known, English is in a fair way to becoming a universal language of science.

Throughout the course of the nineteenth century the number of mathematical journals continued to grow steadily, a process which was accelerated after 1920 with the increase in the number of countries where mathematical studies were being developed, and which culminated, after 1950, in a veritable explosion; so that today there are about 500 mathematical periodicals in the world. To prevent people from drowning in this ocean of information, journals devoted exclusively to listing and summarizing other publications have been created since the last third of the nineteenth century. The most widely disseminated of these journals, the American *Mathematical Reviews*, now extends to almost 4,000 pages a year (with an average of 5 to 10 articles analysed on each page).

Along with the multiplication of periodicals has gone that of expository works, often grouped together in series of monographs (at times quite specialized). Thus it is rare nowadays for a new theory to wait more than 10 years before becoming the subject of instructional expositions.

Towards the middle of the nineteenth century, there was born in German universities the practice of holding "seminars": under the direction of one or more teachers, several mathematicians, local or visiting, among whom are often included students working for doctorates, analyse the state of a problem, or review the most noted new results in the course of periodic sessions during the academic year. Since 1920, this practice has spread all over the world; the expositions which are made at seminars are often disseminated in the form of printed papers, and can thus reach a larger public. The same thing happens with specialized courses taught in many universities and addressed to advanced students.

It has, however, long been evident that in the field of science verbal exchanges are often more productive than the reading of papers. It has been traditional, since the Middle Ages, for students to travel from one university to another, and this tradition is perpetuated in our day, notably in Germany. In addition, invitations to academics to come to work and teach in universities other than their own have become quite frequent.

This need for personal contacts has been institutionalized since 1897 in the International Congress of Mathematicians, which, since 1900, has met every four years (with two interruptions caused by the world wars). The ever larger numbers participating in these Congresses have to some extent reduced their usefulness, and since 1935 more limited gatherings have proliferated: colloquia, symposia, workshops, summer schools, etc., where a more-or-less open-ended range of specialists meet to take stock of their recent discoveries and discuss the problems of the day.

4. Masters and Schools

The spirit of competition, the concern with ranking, the race for records and prizes have never been so furious as in our day: this fact stares you in the face where sport is concerned, and the same spirit extends to intellectual pursuits such as chess or bridge, and even to many others more or less laughable.

In spite of this, we have recently seen passionately advocated the view that all human beings have the same capacity for intellectual creativity, and that the flagrant inequalities we notice in this respect depend only on whether individuals are more or less favoured by education. A curious doctrine, which entails the belief that the brain has a different physiology from other organs, and at which we can only shrug our shoulders when we think of all the children of princes or of millionaires who have remained incurably stupid despite all the care taken to provide them with the best instructors.

We must then admit that, as in all disciplines, ability among mathematicians varies considerably from one individual to another. In the "developed" countries, the teaching of mathematics is assured by a large staff of academics, most of whom have had to obtain a doctorate (or its equivalent) by producing a work of original research, but these are far from being all equally gifted in research capability, since they may lack creative imagination.

The great majority of these works are in fact what we call *trivial*, that is, they are limited to drawing some obvious conclusions from well-known principles. Often the theses of these mathematicians are not even published; inspired by a "supervisor", they reflect the latter's ideas more than those of their authors. And so, once left to their own resources, such people publish articles at rare intervals relating directly to their theses and then very soon cease all original publication. All the same, these people have an indubitable importance: apart from the dominant social rôle which they play in the education of all the scientific élite of a country, it is they who can pick out, in their first years at the university, the particularly gifted students who will be the mathematicians of the next generation. If they are able to keep up to date with advances in their science and to use the knowledge to enrich their teaching, they will awaken hesitant vocations and will direct them to colleagues charged with guiding the first steps of future researchers.

At a higher level we find a far less numerous category (especially in countries like the United States, which need huge teaching staffs), the class of mathematicians capable of going beyond their thesis work, and even of becoming involved in totally different areas; they often remain active in research for thirty years or so, and publish several dozen original papers. In the "developed" countries you can say that, taking the good years with the bad, such a mathematician is born each year for every ten million people, which gives you about 150 active creative mathematicians in a country the size of France, and 600 in the United States or the Soviet Union. It is only they who can usefully take on teaching of the kind needed for graduate courses through which new ideas are spread, and who can effectively advise young mathematicians engaged in research.

Finally, there are the great innovators, whose ideas divert the entire course of the science of their time and which sometimes have repercussions over more than a century. But, as Einstein said to Paul Valéry, "an idea is so rare!" (obviously Einstein was thinking of a *great* idea). We can count half a dozen geniuses of this kind in the eighteenth century, thirty or so in the nineteenth, and nowadays we can reckon that one or two a year arise in the whole world. To work with these great mathematicians is an uplifting and enriching experience; Hardy saw in his association with Littlewood and Ramanujan "the decisive event of his life" ([7], p.88), and I can say as much of my collaboration with the Bourbaki team.

There is no Nobel Prize for mathematicians. But since 1936 the International Congress of Mathematicians, which meets every four years, has awarded at each of its sessions two, three or four medals, known as "Fields medals", to mathematicians, preferably less than 40 years old, whose work is judged the most outstanding by an international committee. Up to now there have been ten Fields medallists from the United States, five French, four English, three Scandinavian, two Russian, two Japanese, one German, one Belgian, one Chinese and one Italian. It is always possible to criticize the way in which these prizes are distributed, or even their principle. I think, however, that nobody can dispute the merit of the medallists, all of whom have had at least one great idea in the course of their career. We should, moreover, add to this list some fifteen mathematicians of comparable excellence but who could not be rewarded on account of the restricted number of medals.

The question of nationality has no greater importance for mathematicians than for research-workers in other sciences; once the linguistic barriers have been overcome (most often through the use of English), a French mathematician will feel much closer, as regards his concepts and his methods, to a Chinese mathematician than to an engineer from his own country.

Natural associations between mathematicians are based much less on countries than on *schools*. While many of these latter are centred on a country, there are countries where several schools flourish, and schools whose influence extends well beyond national frontiers. These are not, of course, entities clearly defined in time and space, but groupings attached to a continuing tradition, with masters in common, and certain preferred subjects and methods.

Before 1800, mathematicians, few in number and scattered, did not have pupils, properly speaking, even though, from the middle of the seventeenth century, communications between them were frequent, and research in mathematics saw almost uninterrupted progress. The first school of mathematics, chronologically speaking, was formed in Paris after the Revolution, thanks to the foundation of the Ecole Polytechnique, which provided a nursery for mathematicians until about 1880, after which date the baton was taken up by the Ecole Normale Supérieure. In Germany, Gauss was still isolated, but the generations which followed created centres of mathematical research in several universities, of which the most important were Berlin and Göttingen. In England, mathematical research in universities entered a phase of lethargy after 1780; it did not wake up until about 1830, with the Cambridge school, which was especially prolific during the nineteenth century in logic and in algebra, and has continued to shine in mathematical physics up to our own day. Italy had only a few isolated mathematicians from 1700 to 1850, but several active schools have developed there since, notably in algebraic geometry, differential geometry and functional analysis.

During the twentieth century, owing to wars and revolutions, schools of mathematics have experienced many vicissitudes. There is first a phenomenon which is difficult to explain in view of the very different social and political conditions in the countries where it occurred. After the 1914–1918 war there suddenly appeared a constellation of mathematicians of the first order in the Soviet Union and in Poland, both of which had up to then produced only a small number of scientists of international renown. The same phenomenon occurred in Japan after the 1939–1945 war, even though the poor and rigid university system there meant that the Japanese school lost some of its best members to the United States.

The factors which can impede mathematical progress are easier to understand. In France the body of young scientists was bled white by the hecatomb of 1914–1918, and the French school became turned in on itself, grouped around its older members. Germany, on the other hand, succeeded better in preserving the lives of its scientists and in maintaining its great tradition of universality, assuring an exceptional degree of influence to its schools of mathematics. Students from many countries went there to be trained, notably young Frenchmen eager to renew ties with traditions forgotten in their own country since the death of Poincaré in 1912. After 1933, however, the flowering of the German and Italian schools was to be brutally arrested by fascism. Only after 1950 were they to be re-established, influenced this time (by a curious reversal of the situation) by French mathematicians of the "Bourbaki" tendency. In Poland the mathematical schools were physically annihilated, since half the mathematians were massacred by the Nazis. They did not recover their standing until after 1970. As for the English and Russian schools, they were able to come through the ordeal without suffering much damage.

Finally, it is appropriate to dwell on the establishment of schools of mathematics in the United States, for this illustrates the difficulty of implanting a tradition of research where one does not exist, even in a country rich in people and in resources. Up until 1870, no creative mathematician of any renown made his presence felt there; the development of a continent left scant room for abstract speculation. After 1880, the first efforts to create centres for mathematics consisted in inviting certain European (especially English) academics to come and teach in the universities, in founding periodicals devoted to mathematics and in sending gifted young students to be grounded in European mathematics. These efforts were crowned with success from about 1900.

The first school to win international notice being that of Chicago, followed between 1915 and 1930 by those of Harvard and Princeton. An unexpected reinforcement came from the mass emigration of European mathematicians driven out by totalitarian régimes. It was they who were to contribute powerfully to the flowering of the very brilliant *home-grown* American schools of our day, which have placed themselves right in the fore by sensational discoveries, in group theory, algebraic topology and differential topology, among others.

The same slow development of schools of mathematics has taken place in India and in Latin America, but with more difficulty because of political and economic upheavals. China too, after 60 years of troubles, seems also to be engaged in this process, which could soon result in the arrival on the scene of some highly talented mathematicians, judging from those who were formerly able to pursue their careers outside their own country. However, many countries in the process of development have not yet succeeded in founding a lasting school. Further, in this as in all the sciences, it is necessary to reach a "critical mass", and except in exceptional cases (the Scandinavian countries since 1900, Hungary between 1900 and 1940) countries which are too small cannot hope to have a true national school, and need to attach themselves somehow or other to those of their more populous neighbours.

The attraction of large and active mathematical schools is easy to understand. The young mathematician on his own can quickly become discouraged by the vast extent of a bibliography in which he is all at sea. In a major centre, listening to his masters and his seniors, as well as to the visitors from abroad who throng in, the apprentice researcher will soon be in a position to distinguish what is essential from what is secondary in the ideas and results which will form the basis of his work. He will be guided towards key works, informed of the great problems of the day and of the methods by which they are attacked, warned against infertile areas, and at times inspired by unexpected connections between his own research and that of his colleagues.

Thanks to these special centres of research and to the network of communications which links them together over the whole planet, it is hardly likely that there could be the same lack of understanding in our day as certain innovators have had to bear in the past. In fact, as soon as an important result is announced, the proof is everywhere, to some extent, busily scrutinized and studied in the months which follow.

II The Nature of Mathematical Problems

1. "Pure" Mathematics and "Applied" Mathematics

"Basic science" and "applied science" are current terms in the language of our day. It seems to be commonly accepted that a science, or a part of a science, is "basic" if its purpose is the understanding of phenomena, "applied" if it has in view the mastery of these phenomena by human beings for their own ends, good or bad. If some people fear "applied" science because of the catastrophes it can bring about, there are few who dispute the value of "basic" science, detached if possible from the use which is made of it.

Ordinarily we do not speak of "basic mathematics", but we think of "pure mathematics" as opposed to "applied mathematics", a fact which could be interpreted as meaning that there is more difference between these two parts of the subject than is the case with other sciences. And indeed, if we analyse a little more closely what is meant by these names, we reach the conclusion that they are actually two *very different* mental processes, which it might be better to call "mathematics" and "applications of mathematics". In fact, when we compare, for example, molecular biology and clinical pathology, we find that they are both concerned with the same objects, cells, considered in the one case with respect to their internal structure, in the other with respect to the total functioning of a bodily organ of which they are the components. By contrast, the "objects" dealt with by mathematicians are not of the same nature at all as those of the engineers and the physicists; the following chapters will be to a large extent devoted to making clear what mathematical "objects" are. We can, however, already say what the Greeks understood clearly, that is, that our senses can never grasp a number or a plane, as distinct from a pile of apples or a wall.

In what then consist the "applications" of mathematics? It seems to me that they can be described in the following way. It is a matter of predicting the behaviour of certain objects in the physical world under given conditions, taking account of the general laws governing this behaviour. A mathematical *model* is constructed of the situation being studied by associating material objects with mathematical objects which are supposed to "represent" them, and the laws which govern the former with the mathematical *relations* between the latter. The original problem is thus *translated* into a mathematical problem. If it can be solved, in a precise or an approximate way, the solution is translated back again, and so the problem is "solved".

An example from today's world is the guidance of artificial satellites and interplanetary spacecrafts. Such an object can be represented by a point in space, to which is added a coefficient representing its mass, that is, a given number, and its position can be plotted by means of its three coordinates in relation to a system of fixed axes (three more numbers); finally the point in time of the observation is marked by a clock, yet another number. The spacecraft is subjected to the gravitational force exercised by the earth, the moon, the sun, and eventually by other planets; in the model these forces are represented by *vectors* whose components are understood as functions of the spacecraft's coordinates. The course of its movement is thus, by virtue of the laws of dynamics, translated into a mathematical model, the solution of a system of differential equations. Mathematicians apply methods which make it possible to give an approximate solution of such a system, that is to say, to make known at each point in time the spacecraft's coordinates, with only a slight margin of error; but before the introduction of computers, it would have taken months or years of "manual" calculations to obtain these figures. In our day we have computers powerful enough to make such calculations almost instantaneously, and so we can predict how fast and in what direction the spacecraft must be launched to make it follow the desired trajectory.

This example is a particularly simple case, because it involves the calculation of only three numbers, which are called the *parameters* of the spacecraft, as functions of a fourth parameter, that of time. There are generally far more "parameters", and the relations between them are far more complicated than Newton's law.

It was a great achievement of seventeenth- and eighteenth-century mathematics to provide mathematical models for the laws of mechanics and for the movements of the planets (known as "celestial mechanics") which accorded admirably with what was observed. Indeed, these models constituted the first really fruitful applications of mathematics. The theorems used in this way belong to what can be called the "elementary" part of the calculus; it called forth the admiration of contemporaries, but now it has lost the glamour of novelty, and is taught in the last year of secondary school or the first two years of university. So it is that mathematicians, as Hardy says, find it "dull, boring and without aesthetic value". All the same, as we shall see in Chapter IV, it is unusual for a mathematical problem, even once it is solved, not to give rise to a host of others. In the case of the differential equations of dynamics, after a period of stagnation a whole new area of mathematics arose with the work of H. Poincaré from 1880 onwards: the subject known as the theory of "dynamical systems", fertile in difficult and profound questions, and one to which numerous mathematicians at present devote themselves more eagerly than ever.

2. Theoretical Physics and Mathematics

The applications of mathematics to physics were extended during the nineteenth century and, further still, in the twentieth century, and their character was altered. When it came to constructing mathematical models for the new theories of hydrodynamics, of elasticity, of electromagnetism, and later of relativity and quantum mechanics, mathematicians found themselves faced with far more formidable problems than those of celestial mechanics, involving notably what are called partial differential equations. One of the simplest examples is the equation of "vibrating strings",

$$\frac{\partial^2 u}{\partial t^2} - c^2 \frac{\partial^2 u}{\partial x^2} = 0,$$

where x is the position of a point on the string, t the time, and $u(x,t)$ the distance by which the point on the string at position x is removed from this position at the time t.

Even though some of these models date back to the eighteenth century, the mathematicians of that era were quite incapable of solving the problems which they posed. It took all the efforts of the analysts of the nineteenth century to obtain significant results of use to physicists. Many of these problems moreover are still only partially solved today, and these continue to give rise to new research and new methods of attack.

We can therefore say unhesitatingly that it was the requirements of physics which led mathematicians to create a new branch of their science, which we call functional analysis. This is now a considerable area in mathematics, and one which is continually progressing and ramifying. It is noteworthy that it is sometimes by transposing into mathematical models certain concepts appertaining to physics (energy, the principle of "minimum action", etc.) that general methods have been obtained for solving certain equations in functional analysis.

Again, more recently, the use of probability theory in statistics and in physics has led to many works on this theory; and in the same way operations research and the theory of automata have stimulated the study of certain branches of algebra.

3. Applications of Mathematics in the Classical Era

Newtonian mechanics and its resounding successes in astronomy, notably Clairaut's prediction of the return of Halley's comet, accurate to within one month, made a strong impression on all the intellectual circles of Europe, even on men of letters such as Voltaire, who understood absolutely nothing of mathematics[1]. As for eighteenth-century mathematicians such as Clairaut,

---

[1] He wrote somewhere that he could never understand why the sine of an angle is not proportional to the angle.

d'Alembert, and Laplace, these remarkable applications of mathematics, followed by many others, led them to think that the essential goal of mathematical research was to supply models for mechanics and physics; any branch of mathematics which did not meet these conditions being deemed futile and negligible. When Euler communicated to Clairaut his results in number theory, on which he said he had been working for fourteen years, Clairaut only remarked, "That must be a very thorny subject", and switched forthwith to the subject which interested him, the calculation of the perturbations of the planets ([6], p.129).

Before the seventeenth century the applications of mathematics were far from having the quasi-magical character which awoke admiration for Newton's mechanics. In fact, there were only four sciences in which mathematicians had been able to supply models making possible a *rational* explanation of phenomena. These were all due to Greek mathematicians, after whose day no new idea arose before the sixteenth century. They were: the optics of mirrors (or "catoptrics"), statics, the equilibrium of floating bodies, all of which seem to have originated no earlier than the fourth century B.C., and astronomy, the first steps in which were recorded in the sixth century.

These applications are based on geometry alone, and are seen as annexes to that subject[2]. For lack of a mathematical tool which would have enabled them to describe an arbitrary motion, such as became available only with the calculus, the Greeks, in order to "keep track" of the movements of the planets in the celestial sphere, began with an *a priori* idea, that is, that only uniform rotations around an axis (or, in a plane, around a point) were acceptable for heavenly bodies to which they ascribed a superior "perfection", no doubt a residue of theological concepts concerning the "divinity" of these bodies. This led them, as is well known, to imagine a system of spheres, each moving in relation to others, or of "epicycles" (circles whose centres describe other circles), a system which became more and more complicated to fit the observations as the latter became more precise.

But before making this attempt, doomed eventually to failure, the Greeks managed to form an idea, approximate as yet but still basically right, of the shape and dimensions of the earth and of the two heavenly bodies whose diameter is clearly appreciable to the naked eye, the moon and the sun. For this they used only their geometry, applied in an original way to observations which were simple, but chosen with rare ingenuity. I should like to summarize here these first achievements, so different in spirit from the attitudes of other peoples towards the same phenomena. Even the Babylonians, whose astronomy was the most developed, went no further than noting the movements of the heavenly bodies (their rising and setting, their conjunctions and eclipses, etc.), and discovering the multiple periodicities affecting them; but no sign is to be found in their thought of a geometric model designed to *explain* them.

---

[2] In order to work out his geometrical theorems for areas, volumes and barycentres, Archimedes explains, in a little work rediscovered in about 1900, how he splits his figures into "slices", which he compares in terms of nominal "weight".

We must remember firstly that the only instruments of measurement for astronomic observations available in the Ancient World and in the Middle Ages were rulers of varying length, known as "alidades", pivoting around the centre of a graduated circle, and making it possible, by looking through holes placed at the far ends of the alidade, to determine the *angle* between the straight lines leading to two points on the celestial sphere, with a varying degree of precison which might be of the order of a minute on an arc for the best instruments. Every model had therefore to be reduced for them to the measurement of angles.

The first step was taken by the Pythagoreans of the sixth century B.C., who were convinced that the earth, the moon and the sun were spherical bodies and that the phases of the moon resulted from the variations in the portion lit by the sun and were caused by the movements of the sun and the moon, respectively, around the earth, thought to be fixed. They understood that the eclipses of the moon were caused by its passing through the cone of the earth's shadow (Figure 1); the fact that we see the trace of this cone on the moon as the arc of a circle (Figure 2) is explicable only on the assumption that the earth is a sphere.

Another observation taken of the moon enabled them to find out that the distance ES is much greater than EM[3]. In fact, during the first quarter, the plane separating the illuminated part of the moon from the part in shadow passes through E, and the straight line EM is therefore perpendicular to ES (Figure 3); hence

$$\frac{EM}{ES} = \cos\alpha.$$

The angle α is very near to 90° (it differs by less than a minute), and the ratio EM/ES is thus of the order of 0.003. As sightings on the sun are difficult, the Greeks thought that α was near to 87°, and therefore obtained too small a value for ES.

Thus we make only a very small error if we consider that at all points of the earth's surface all the rays coming from the centre of the sun's disc are parallel. This enabled Eratosthenes, in about 300 B.C, to obtain a first approximation of the earth's radius, by the following chain of reasoning. In Egypt, Alexandria and Syena (now Aswan) are more or less on the same meridian (Figure 4). On the day of the summer solstice, the sun at midday is exactly overhead at Syena, and at the same time its rays make an angle β with the vertical at Alexandria. As the angle β is that formed by the radii joining the centre of the earth O to Alexandria and to Syena, the arc of the meridian between these two towns is $d = R\beta$ (β in radians), whence $R = d/\beta$, and Eratosthenes had a rough estimate of the distance d.

---

[3] In Figures 1 to 3, E, S, M are points which can be taken either at the centres of the three bodies or on their surfaces; the errors made by varying these points are less than the errors in the measurement of angles which resulted from the imperfection of the Greeks' instruments.

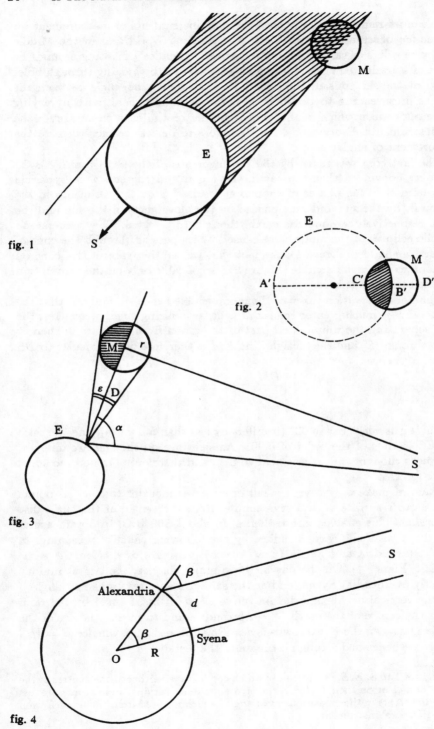

fig. 1

E

S

fig. 2

E

A'   C'   D'
              B'   M

fig. 3

M   r

ε   D

E   α

S

fig. 4

S

β

Alexandria

d

β

O   R   Syena

Given the value of R it is possible to deduce the distance EM of the earth from the moon if we know the ratio $A'B'/C'D'$ (Figure 2) of the radii of the earth and of the moon. The Greeks judged that this ratio is near to 3 by observing the curvature of the earth's shadow on the moon, whereas in fact the ratio is 3.7. If we then measure the angle ε which the lunar disc subtends, we have (Figure 3) that $D = 2r/\varepsilon$, where r is the lunar radius.

It will be observed that the applications of mathematics which we have just described, as also the celestial mechanics of Newton and his eighteenth-century successors, are concerned with a "basic" science which in itself had no immediate "use" before the invention of artificial satellites. The universal admiration which this science has, even so, always been accorded is an example of a tradition unbroken since the Greeks, that of attributing value "to sciences which are not concerned with utility or profit", as Aristotle puts it ($An.\ Pr.$ 981^b). We touch here on one of the characteristics of human beings, $curiosity$, already so evident in young children, even if many adults seem to have lost it and to be sunk in routine and intellectual inertia.

Of course, the Greeks would never have thought of dissociating the properties of numbers or of geometrical figures from their applications in astronomy. By contrast, whole numbers, from the time of the Pythagoreans of the sixth century B.C., have tended to be invested with a kind of mysticism, still very evident in Plato, and which has been perpetuated until our own day in the vagaries of "numerology".

4. The Utilitarian Attacks

It was in the circle around d'Alembert, with Buffon and especially Diderot, that the dispute about the proper place of mathematics was to be extended much further, to the point of making its applications to other sciences the only objective which mathematicians ought to have in view. For Diderot, mathematics had had its day: it added nothing to experience, and merely "interposed a veil between nature and the people", instead of "making philosophy popular". We see that the principles of the "cultural revolution" do not date from yesterday! Diderot, unlike Voltaire, had a smattering of mathematics, but he must have realized that he would never produce an original work. Should we attribute his attacks to this? As Fontenelle was already writing in 1699: "We commonly call things useless which we do not understand. It is a kind of revenge, and since mathematics and physics are generally not understood, they are declared useless."

To be fair to Diderot and to those who, up to our our own day, have repeated his invectives against mathematics, it must be recognized that until the beginning of the twentieth century, mechanics and theoretical physics had scarcely had any influence on "useful" technological inventions: the "simple machines" (lever, pulley, screw, etc.), go back to remote antiquity; the steam engine preceded the law of thermodynamics, and even the combustion engine and aviation resulted from "pottering about" rather than from theories. It is only in our day that technology can no longer do without the results of "basic science", and hence of mathematics. Thus the revolutionaries who thought they could restrict the work of mathematicians to what could be "popular", in Diderot's sense, had to go quickly into reverse.

It is still hard to understand the spleen which erupts in pamphlets from time to time, constantly attacking "pure" mathematics. There are plenty of other disciplines, such as cosmology, prehistory and archaeology, which clearly have no "use", but nobody attacks these. On the contrary, it is easy to rouse public opinion in their support when a government reduces the funds allocated to them. Must we again invoke Fontenelle's dictum?

5. Fashionable Dogmas

Those who despise "useless" mathematics have to recognize that, amid the huge production of contemporary mathematics, a considerable part has visibly no "application" conceivable in our day. They avoid this difficulty by affirming forcefully that even these parts will one day become "useful". As such an affirmation refers to an indeterminate future, this is nothing but dogma, as impossible to refute as to prove. The crowning argument is that Kepler found the mathematical model of the movement of the planets around the sun in the theory of conics, developed 1800 years before by Apollonius without the least intention of "applying" it to anything whatsoever. More recently, it has been possible to add two similar cases: general relativity, which found its model in Riemannian geometry, devised sixty years earlier, and the structure of atoms and of "elementary particles", in which group theory and Hilbert spaces play important parts.

But here are three of the most remarkable theorems of number theory:

I (Euclid, around 300 B.C.) There exists an infinity of prime numbers (or, in the original formulation, *each* natural number is smaller than some prime number).

II (Lagrange, 1770) *Each* natural number can be written as the sum of one, two, three or four squares of natural numbers (for example: $7 = 2^2 + 1^2 + 1^2 + 1^2$).

III (Vinogradov, 1937) There exists a (very large) natural number N such that *each* odd number greater than N is a prime number or the sum of three odd prime numbers[4].

---

[4] As an example of such a decomposition: $33 = 23 + 7 + 3 = 11 + 11 + 11$. It is conjectured that this decomposition works for *any* odd number; we should therefore verify it for whole numbers smaller than N, but N is too large for this to be possible, even with the most powerful computers. If any reader imagines that such statements are rare in mathematics, I would have him consult the 1289 pages of L.E. Dickson's work, *History of the Theory of Numbers* (2 vols., Carnegie Institute, Washington, D.C., 1920). They describe exhaustively all the papers relating to number theory from the beginnings to 1918. An equally large book would be needed to take account of all that has been done since then in this branch of mathematics.

Each of these results states a property pertaining to an infinity of natural numbers. If we weakened the statements by limiting ourselves to natural numbers $< 10^{1000}$, for example, we should still have theorems, but they would not be of much interest to most mathematicians. It is difficult to imagine that a physical theory would need to consider an infinite set of numbers.

There are other opinions which mathematicians find astonishing, such as are formulated by certain historians of science. These people do not consider adequate those works on the history of mathematics which describe the ideas of the past and seek to understand their interconnections and the influence some have had on others: it is also necessary, according to them, to "explain" why mathematicians chose such or such a direction in research, and how they arrived at their results. I confess I do not understand what that can mean: what goes on in a creative mind never has a rational "explanation", in mathematics any more than elsewhere. All that we know is that it entails periods, sometimes long ones, of unproductive investigations, followed by sudden "illuminations", and a "formalizing" of what these have revealed. All this has been well described by Poincaré from his own experience [9].

As for the origins of the problems which a mathematician sets out to solve, they must almost always be sought in his contacts with other mathematicians, whether through reading classic works or explanatory texts, or through conversation or exchange of letters, or again through listening to talks. But this is not enough for the historians of science, who are critical of the concept that mathematicians form of their own science, and undertake to bring to bear their universal "explanation", the influence of social environment. This is known to be a dogma proclaimed by many intellectuals in the desire to minimize the contribution and the originality of individuals and to combat what they call "élitism". I am not competent to judge of the validity of this dogma in other sciences, although I cannot see how the societies in which they lived can have influenced the discoveries of Newton or of Einstein. But in mathematics, not counting those parts which serve as models for the other sciences, the dogma seems to me absolutely absurd, in view of what is known of the way in which mathematicians of the past have worked and of the behaviour of our contemporaries. How can it be thought that social environments as diverse as the Alexandria of the Ptolemys for Euclid, the Paris of Louis XV or the Berlin of Frederick II for Lagrange, or the Soviet Union under Stalin for Vinogradov, could have had anything whatever in common with their theorems in number theory cited above?

There is another aspect of this dogma, which claims to see in the motivation of a researcher in science the desire to be "useful" to society. I shall confine myself to quoting what Hardy says about this ([7], p.19): "If a mathematician, or a chemist, or even a physiologist, were to tell me that the driving force in his work had been the desire to benefit humanity, then I should not believe him (nor should I think the better of him if I did)." I am convinced that this is what most mathematicians think, even if they do not say so openly for fear of being put to shame by the reigning "intelligentsia".

6. Conclusions

Let us summarize this long discussion in a few sentences.

I. There is a whole *important* area of mathematics which arose in order to supply models for other sciences, and there is no question of minimizing this. All the same, it constitutes certainly no more than 30% to 40% of the set of contemporary mathematics, as can easily be recognized if we run through the monthly publication *Mathematical Reviews*, which gives summary analyses of everything that is published in mathematics and in the most important of its applications.

II. We can say with Hardy ([7], p.20) that the main reason which impels a mathematician to do research is intellectual curiosity, the attraction of enigmas, the need to know the truth: "the problem is there, you *must* solve it", said Hilbert. We shall see in Chapter IV how this insatiable passion has led to the extraordinary burgeoning of questions which mathematicians never cease to put to themselves.

After this, of course, come other motives which are not in the least dishonourable and which mathematicians share with many other people, such as the wish to leave something lasting behind them, and also surely the ambition to be esteemed in their lifetime for what they have accomplished.

III. The mathematician is very well aware that this esteem can come to him only from his peers. He lives in a closed circle, almost without communication with the outside world, and it is from within this circle that his problems come to him, with the exception of those parts which serve as models for other sciences.

IV. However high an opinion a mathematician may have of his own works, it is the judgement of his peers which will establish their value. It is obvious that his results must be "non-trivial": they will be the more admired the more difficult they were to achieve, and in so far as the problems solved had often been attacked unsuccessfully before. Beyond this, there is no criterion for appreciation which does not vary from one epoch to another and from one mathematician to another. Terms such as "generality" or "depth" are used, but they do not always have the same sense for those who use them. There are also fashions and fads, which temporarily boost some areas of mathematics at the expense of others. These divergences in taste recall the quarrels aroused by works of art, and it is a fact that mathematicians often discuss among themselves whether a theorem is more or less "beautiful". This never fails to surprise practitioners of other sciences: for them the sole criterion is the "truth" of a theory or a formula; that is to say, the way in which it accounts more or less satisfactorily for observed phenomena. In mathematics, all the results are "true", in the sense that they have been proved according to the accepted logical rules (cf. Chapters III and VI), an unproved assertion having no place in mathematics. Other criteria are therefore necessary to evaluate mathematical work, and these are unavoidably subjective, a fact which makes some people say that mathematics is much more an art than a science.

III Objects and Methods in Classical Mathematics

While all ancient civilisations, in order to satisfy the needs of daily life, had to develop procedures of arithmetical calculation and spatial measurement, only the Greeks, from the sixth century B.C., thought of analysing the chain of reasoning behind these procedures, and thus created an entirely new mode of thinking. In this chapter we shall try to make plain the essential features of Greek mathematics and the unsuspected developments, so extraordinarily fruitful, which mathematicians brought to them between the Renaissance and the end of the eighteenth century.

In §§2 and 3 we shall focus on the two fundamental characteristics of Greek mathematics:

1) The idea of *proof*, by a succession of logical inferences from unproved propositions, axioms and postulates. It must be emphasized that this idea could only be brought into play thanks to the skill acquired in the manipulation of logic by those nursed in the Greek schools of philosophy. A particularly striking example is supplied by the principle of "proof by contradiction", a tool refined by logicians which became one of the pillars of mathematical reasoning.

2) The objects with which mathematicians are concerned have the same names as those which are used in practical calculations: numbers, geometrical figures and magnitudes. But, from the time of Plato, mathematicians have been aware that under these names they are reasoning about entirely different entities, *immaterial* entities, obtained "by abstraction" from objects which are accessible to our senses, but which are only "pictures" of the former.

We show in §4, which is about diagrams in geometry, how the properties attributed by the axioms to the "abstract" objects of geometry make them profoundly different from their "pictures", and what difficulties arise because of this in finding an appropriate vocabulary to *define* these objects. So that we may be clear about these ideas, we shall not stop here to trace in detail the historical vicissitudes of these concepts, which we shall come back to in Chapter VI. Instead we shall pass straight away to the step taken by Pasch and Hilbert at the end of the nineteenth century; taking up Euclid's axiomatic method in the original spirit, but filling in his lacunae, they were able to escape once and for all from these difficulties, by stating that it is the axioms governing mathematical objects which *define* these objects.

In the same way, §5 is devoted to mathematical objects whose "pictures" are the numbers and magnitudes of the reality perceived by our senses. The "abstract" character of the idea of a whole number was always present in Greek arithmetic, and Euclid's exposition of the divisibility of whole numbers and of prime numbers is still substantially that which is taught today. Although, unlike his geometry, it is not put in the form of an axiomatic theory, very little needs to be added to give it that form (Chapter VI).

By contrast, the discovery of incommensurable magnitudes brought about a crisis in the Greek concept of the measurement of magnitudes. It seems in fact that formerly the Pythagoreans had always accepted that once a unit was chosen for some kind of magnitude, every magnitude of the same kind was "commensurable" with this unit – we would say that its measure is a rational number. To overcome this difficulty, the Greeks created new mathematical objects, namely the *ratios* between magnitudes of the same kind. These were defined axiomatically in such a way that the ratios between magnitudes of the same kind to a unit chosen once for them all constitute a *part* of what we call the set of *positive real numbers*. This part contains the rational numbers and certain irrational numbers, but the texts do not allow us to specify exactly of what it consists.

No doubt for philosophical reasons, the mathematicians of Plato's school observed taboos in the manipulation of the three kinds of geometric magnitude: length, area and volume. For example, you could not add the measures of a length and of an area, and the product of the measures of two lengths was the measure of an area (or that of three, the measure of a volume), and not the measure of a length. Although geometry might accommodate itself to these restrictions, they rendered impossible such algebraic calculations as we practise on the real numbers. The authority of Descartes was needed to sanction the adoption of this kind of calculation, although certain mathematicians had already suggested them in previous centuries. From this time on, the "ratios" between magnitudes of the same kind were identified with the real numbers, without its being necessary to specify the kind under consideration.

In §§7 and 8, we show how, together with the invention of a convenient notation in the course of the Middle Ages and the Renaissance, this reform made possible not only the development of algebra, but also the essential invention of the *method of coordinates*, thus supplying on the one hand an algebraic model of Euclidean geometry, and making possible on the other the emergence of the general idea of the real *function* of a real variable, an idea which remained unknown to the Greeks.

Finally, §§6 and 9 introduce two of the most fundamental ideas of mathematics, that of *approximation* and that of *limit*, which is derived from it. The Greek mathematicians usually solved algebraic problems by geometric "constructions"; for example, Euclid gives the construction of the square root of a "ratio" by the intersection of a circle and a straight line, and Menaechmus in the same way constructs a cube root by the intersection of two conics. Yet we also find in Euclid another idea for defining the measure of the area of a plane, non-polygonal figure: he inserts the figure between two sequences of polygons the difference in whose areas tends to zero. This idea was used repeatedly by Archimedes; it was generalized in the seventeenth century, and it makes it possible, for instance, to prove the existence of nth roots for $n \geq 4$, something which the Greeks could not achieve by means of geometric constructions. It is of course necessary, in order to justify these procedures, to introduce an axiom which remained implicit until the nineteenth century, when it was eventually made explicit by Cauchy under the name of "the Axiom of Nested Intervals". This axiom, in conjunction with those of Euclid, completes the axiomatic definition of the set of *all* real numbers. It provides a solid base for the *calculus*, a subject which was invented in the seventeenth century and was destined to become the most powerful tool of pure mathematics and of its applications.

1. The Birth of Pre-Mathematical Ideas

In the society of our day, the ideas of number and of measurement are acquired in infancy. From the age of twelve or thirteen they seem so "natural" that their use becomes automatic. Piaget, however, showed by experiment that, whilst the idea of an "arbitrary" natural number may be grasped very early, it is otherwise with certain types of magnitude such as volume or weight; children up to the age of twelve still have conceptual difficulties in comparing two magnitudes of the same kind. Ethnologists for their part have come across primitive societies among whom whole numbers in excess of some few units have no name, and *a fortiori* cannot be used in calculations.

The texts which have come down to us from the earliest oriental civilisations, in Egypt or in Babylonia, are too fragmentary to allow us to follow the way in which an arithmetic or a rudimentary geometry came to be constituted. They appear fully developed from the second millennium B.C. Of course we are not dealing here with abstract speculations, but with recipes, handed down by castes of specialized scribes, and designed to regulate the practical problems posed by an already highly structured agrarian society: problems of exchange, rents, litigation, division of property[1].

It is not necessary for our purpose to enter in detail into the problems solved in the documents which have come down to us. Let us say only that in arithmetic they show a knowledge of fractions, of arithmetical progressions and perhaps of geometrical progressions, and of the "rule of three". Among the Babylonians we even find solutions of problems equivalent to a quadratic equation. For example, there is a tablet showing the diagram of a square, with the following text: "I have added the length of the side of the square to its surface area and I find 3/4; what is the side?" The scribe is thinking of the equation which we would write as

---

[1] The Greek authors emphasize the fact that, since the flooding of the Nile altered the shape of the fields, it was necessary to have recourse to a man of special skills, a scribe who held formulae in trust, in order that each person should recover, after the flooding, a plot of land equivalent to what he possessed before.

$$x^2 + x = \frac{3}{4}$$

and which the scribe solves in the same manner as we would: he adds 1/4 to both sides of the equation, and finds that the square of $x + 1/2$ is 1, from which he concludes that $x = 1/2$.

In the field of plane geometry, figures such as rectangles, triangles, trapeziums, the right angle and the circle, were known, possibly in conjunction with the use of tools, such as the potter's wheel, the surveyor's tracing-line, and the mason's optical square. The idea of similarity is found among the Babylonians, one of whose texts states that where we have a staircase the ratio between the height and the width of a single step is the same as that between the total height of the staircase and its horizontal projection (Figure 5). On the other hand, the Greeks attributed to Thales a procedure for measuring the height of a pyramid which was no doubt known to the Egyptians: we observe the length of its shadow, the ratio between the pyramid's height and this length being the same as that between the height of a stick and the length of its shadow (Figure 6). As for three-dimensional geometry, the monuments which survive are witness to empirical knowledge on the part of architects and masons about which it is difficult to be precise.

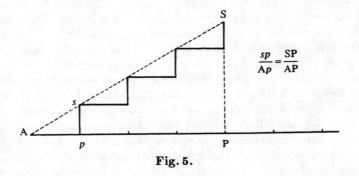

$$\frac{sp}{Ap} = \frac{SP}{AP}$$

Fig. 5.

The concepts which arise are never concerned with any but *concrete* objects: the enumeration of objects in a pile, the measurement of magnitudes capable of being added or subtracted, such as length, area, volume, weight, angle, for each of which a unit is taken and often its multiples or sub-multiples.

The recipes are concerned with examples where the data are *specified*; they are procedures in calculation which are not given a general justification, such as evaluation of areas of figures whose forms and dimensions are known, for example, isosceles triangles, rectangles, trapeziums or circles[2]. Naturally, we do not find *formulae* in the sense in which we understand the word: formulae which are valid for arbitrary or for unspecified data. The general nature of

---

[2] For certain polygons, these formulae give only an approximate value for the area; for the circle, on the other hand, certain Egyptian texts give the good approximate value for π of 3.16.

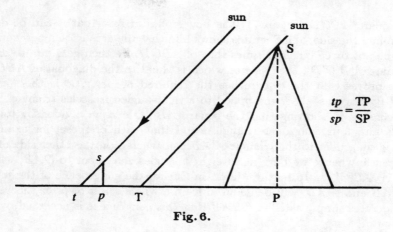

$$\frac{tp}{sp} = \frac{\text{TP}}{\text{SP}}$$

Fig. 6.

the procedure of calculation can only be guessed when a series of examples is given whose data vary.

2. The Idea of Proof

There are no documents left to us from any ancient civilization prior to the fragments of Greek authors of the seventh and sixth centuries B.C. which enable us to discern examples of what we call *logical deduction*: that is, sequences of *inferences* – later codified as syllogisms[3] – which *constrain* an interlocutor to assent to an assertion Q once he has assented to another assertion P[4]. It is known that, from the fifth century B.C., Greek thinkers were past masters in the art of ordering an argument as a succession of logical deductions, as is shown in the fragments of the works of the sophists as well as in Plato's dialogues. They had discovered that these reasonings could take any human activity as their object and in particular the recipes of arithmetic and geometry, most of which no doubt came from the Egyptian and Babylonian civilizations. These were to become the *proofs* linking the theorems together.

Tradition, known only through later texts, has the first theorems dating from Thales, at the end of the seventh century B.C., but the proofs are not known, supposing that they existed. It is, however, accepted that the theorems of the Pythagoreans – among others, of course, the one known as "Pythagoras' theorem" – were accompanied by proofs, even though here again their nature is still unknown. The first texts containing proofs are found only among the writings of Plato and Aristotle.

In the dialogue known as the *Meno*, Socrates wants a young uneducated slave to find out how to construct a square whose area is twice that of a

---

[3] Some of the various sorts of syllogism thought up by the logicians have never been used by mathematicians.

[4] The use of letters to designate unspecified propositions goes back to Aristotle.

given square $ABCD$ (Figure 7). The boy replies at first that it can be done by doubling the side, and Socrates shows him that the area of this new square would be not twice but four times that of $ABCD$. He then gets him to draw the square $A'B'C'D'$, each side of which is equal to the diagonal of $ABCD$, and he proves that this square has the required property. It is the special case of Pythagoras' theorem applied to a right-angled isosceles triangle. The proof consists in showing that the square $ABCD$ can be divided by its diagonals into four congruent triangles, and that each of these, for example OAB, is congruent to the triangle $A'AB$ constructed on the other side of its hypotenuse; whence we have in all eight triangles congruent to OAB, whose union is $A'B'C'D'$. To be sure, later, in Euclid, the congruence of the triangles OAB and $A'AB$, will be deduced from a whole series of prior theorems. Here Socrates simply satisfies himself that this congruence is accepted by his interlocutor.

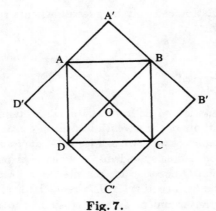

Fig. 7.

The second proof, recorded by Aristotle as coming from the Pythagoreans, is concerned with the same figure of a right-angled isosceles triangle, and constitutes the first known example of proof by "contradiction", which was to become an essential tool of mathematics[5]. It is also the first example of an assertion of *impossibility*. The theorem states that in a right-angled isosceles triangle the ratio between the hypotenuse and one arm of the right angle *cannot be a fraction p/q* (where p and q are whole numbers). In fact, by Pythagoras' theorem, such a fraction would have the property that $(p/q)^2 = 2$. One can suppose the problem "reduced" to the case where p and q *are not both even*: if they are we can divide them by two without altering the value of p/q, and, by doing this several times if necessary, arrive at the "reduced" case. Now $p^2 = 2q^2$, and since the square $(2n+1)^2 = 4(n^2+n)+1$ of an odd number is odd, p is necessarily *even*, say $p = 2p'$. Now $4p'^2 = 2q^2$, whence $q^2 = 2p'^2$, and so q too is necessarily *even*: we have reached an absurd conclusion, and

---

therefore the hypothesis from which we started is untenable. Here the points of departure from which inferences are made are the elementary properties of odd and even numbers, something for which the Pythagoreans seem to have had a predilection. The same predilection is to be found in the ninth book of Euclid's *Elements*, even though by this time the subject had become "trivial".

3. Axioms and Definitions

The way in which a proof must proceed by way of a series of inferences is very well described by Plato in a famous passage of the *Republic* (VI, 510, c, d):

Those who work in geometry and arithmetic ... begin by postulating odd and even numbers, or the various figures and the three kinds of angle [...] These data they take as known; and, having adopted them as assumptions, they do not feel called upon to give any account of them to themselves or to anyone else, but treat them as self-evident. Then, starting from these assumptions, they go on until they arrive, by a series of consistent steps, at all the conclusions they set out to investigate.

This method has remained the same for mathematicians of all time. But reflections on the nature of the "hypotheses" of which Plato speaks (it will be noted that he never uses the word "truth") and on the entities to which they are applied have never ceased since then to occupy the minds of mathematicians and of philosophers. This is what is known as the problem of the "foundations of mathematics".

Many of Plato's dialogues have as their theme the attempt to elucidate "abstract" words used in current speech without any clear conception of what they mean, such as beauty, courage, love, piety, justice, virtue, etc. In the same way mathematicians were to have to sort out what they meant by the words *figure, position, magnitude, quantity, measure*, which are the basic notions appearing in the "hypotheses". Let us say straight away that they were not to succeed completely in this until the end of the nineteenth century, when they reached the completion of a long history some of whose most striking episodes we shall retrace in this and the following chapters.

If these words designated concepts belonging to sensory experience, they would be no more problematic than this experience itself. But at the end of the passage from the *Republic* cited above, Plato takes care to explain immediately (VI, 510, d) that mathematicians

make use of visible figures, and discourse about them, though what they really have in mind is the originals of which these figures are images ... . The student is seeking to behold those forms which only thought can apprehend.

For further illustration we need look no further than the scene in the *Meno* to which Plato is obviously alluding here, to see that what Socrates states has nothing to do with the figures which he has traced – probably in the sand.

He would indeed have difficulty in proving "experimentally" the congruence of the triangles OAB and $A'AB$.

Plato associates with his descriptions considerations relating to his theory of Ideas, which do not concern us here. Yet Aristotle, who does not accept this theory, nevertheless confirms the preceding passage in the *Republic*:

The investigations of mathematicians, he writes (Metaph. $K3, 1062^a$ $20 -^b 3$), *have to do with things reached through abstraction* (ἐκ ἀφαιρέσεως), *for they study them after eliminating all sense data, such as weight, lightness, hardness, etc., retaining nothing but quantity and continuity, this latter being possible to conceive in one, two or three ways* [6].

In this way were introduced the first "mathematical objects". But understanding what they are and what can legitimately be said about them was to entail innumerable difficulties right until the end of the nineteenth century.

While we have an appreciable portion of the works of Plato and Aristotle, hardly anything has come down to us of the writings of mathematicians before Euclid. Hence it is only from the latter's *Elements* that we can obtain fairly precise information about the concepts of the Greek mathematicians of the fifth and fourth centuries; although this work seems to be a compilation drawn from several others from quite different epochs.

The "hypotheses" which form the point of departure of the process described by Plato are for the most part themselves deduced from earlier "hypotheses". For example, the reasoning of Socrates in the *Meno* is a particular case of a theorem of Euclid's (Book I, 34) proving the congruence of the two triangles ABD, BCD in a parallelogram $ABCD$ (Figure 8). For this he uses a "case of the congruence of triangles" (Book I, 26), noting that the side BD is common to the two triangles and that we have equality of the angles $\widehat{ABD} = \widehat{BDC}$ and $\widehat{ADB} = \widehat{DBC}$ by Book I, 29.

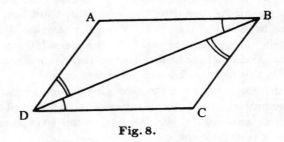

Fig. 8.

It is certain that the idea of "ordering" the theorems of geometry by a succession of proofs came very early, and later commentators cite "Elements" earlier than Euclid's of which nothing has come down to us. But obviously this regression of hypotheses cannot continue indefinitely, and it must come

---

[6] An obvious allusion to the three dimensions of space.

up against "hypotheses" which are *unproved*. It is well known that this is just what we find at the beginning of Euclid's *Elements*, after a certain number of "definitions". It is normal today to group these "hypotheses" under the name of *axioms* of the geometry: this seems to me a mistake, since they do not all relate to the same objects. Euclid already knows this, since he divides them into "requirements" and "common notions". The "requirements" – also called *postulates* – are the properties of plane geometry, while the "common notions" – which at the end of the classical era came to be called *"axioms"* – concern all kinds of "magnitude" (see below, §5).

4. Geometry, from Euclid to Hilbert

It is in Euclid then that we see developed for the first time according to the deductive method the properties of "mathematical objects" conceived in the spirit of Plato and Aristotle. His "definitions" in Books I to VI enumerated those among these objects which belong to plane geometry: point, straight line, angle, circle, polygon (apart from triangles and quadrilaterals only regular polygons are studied in any detail). Euclid leaves no doubt as to the fact that these objects are not accessible to our senses; the two first "definitions" state that "the point is without extension" and that "the line is without thickness"[7]. What is more characteristic are the two "requirements":

1) that *any two points can be joined* by a segment of a straight line;

and

2) that a segment of a straight line can be *indefinitely prolonged* in both directions.

These properties are constantly made use of, but would be absurd if they referred to material "straight lines"!

It is from these "definitions", "requirements" and "common notions" that Euclid claims to prove the sequence of his theorems. What continues to surprise us a little is that each of these is accompanied by a diagram. It might be thought that this is merely an aid to make it easier to follow the proof, and it has been said that the art of geometry is to reason well from false diagrams. But it can immediately be seen that some of these figures play a far more essential role, and that we are not far from what Indian or Chinese geometers

---

[7] These are in fact only unusable "pseudo-definitions". The definition of a word must bring in words defined earlier, and the word defined serves as an abbreviation; when we want to use it we must "always mentally substitute the definition in the place of what is defined", as Pascal expresses it, taking up a precept stated by Aristotle (Top., VI, 4). Now we *never* see Euclid "substituting" these two "definitions" for the words "point" and "line"; we can therefore consider that as far as he is concerned these words *are not defined*.

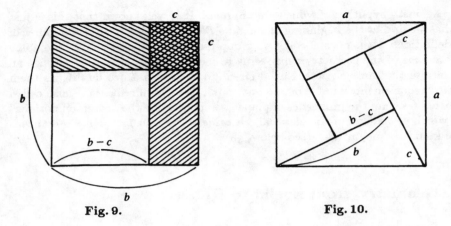

Fig. 9.   Fig. 10.

do when they draw a diagram and then merely say "see" by way of proof[8]. For example, Euclid takes for granted many times (Book III, 17; Book VI, 13) that a straight line containing a point lying inside a circle intersects this circle; in an analogous way, if a circle C has one point inside and one point outside a circle C', then C and C' intersect (Book I, 1 and 22); none of these facts follows from his "requirements". The recourse to "visible" objects is even clearer in Book III, 8, where Euclid studies the segments of a straight line joining a point outside a circle to a point on the circle (Figure 11), and distinguishes on this circle the "convex circumference" (in relation to the exterior point) from the "concave circumference" (notions which he would indeed have difficulty in defining for the "absolute figures" of Plato). And what are we to say of the definitions given in Book XI, on geometry in space, where he treats of "surfaces described" by a straight line or a semi-circle which "turns" around a "fixed" straight line? These examples could be multiplied; they show the difficulties he would have had to surmount to create a vocabulary adapted to the nature of objects only "visible to thought" and to expound their properties in conformity with this nature, that is, *without diagrams.*

Further, Euclid has recourse to "evidence" based on the study of a drawing in only a small number of cases, owing to the restricted number of "diagrams" to which he limits himself. His successors in the ancient world, as we shall see later (§8), would enrich the domain of geometry by discovering and studying new curves and surfaces, thus preparing the ground for the leap forward in mathematics after the Renaissance. But if we look at it closely we see that this progress is accompanied, far more than in Euclid's case, by recourse to a large number of properties which are not formulated, but only suggested by a more or less accurate drawing.

---

[8] In Figure 9, it can be "seen" that $(b - c)^2 = b^2 - 2bc + c^2$, and, in Figure 10, that in the right-angled triangle of hypotenuse a and sides b and c, we have $a^2 = (b - c)^2 + 2bc = b^2 + c^2$, according to Figure 9, which "proves" the theorem of Pythagoras.

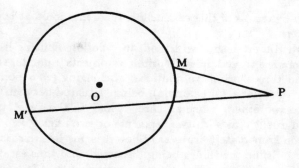

Fig. 11.

It is, however, only our training in the rigours of modern axiomatics which makes us notice these imperfections. Apart from the thorny question of the parallel postulate – the fifth of Euclid's "requirements", to which we shall return (Chapter VI) —, it does not seem that there was much criticism of this kind of the text of Euclid or of the work of his successors before the sixteenth century. Long-established custom seems to have dulled in geometers the awareness of the temerity represented by the transition from the realm of physical objects to objects of which these are only crude images. Given that this awareness is so clear in Plato and Aristotle, it is surprising to see thinkers as profound as Descartes and Pascal – men who did not hesitate to make a frontal attack on scholasticism – proclaiming vigorously the "evident truth" of the axioms of geometry! They are, moreover, only expressing a state of mind common to all the mathematicians of their time, and which in the following century would simply be accentuated, since we see Gauss and Cauchy still competing to vaunt "geometric rigour" as a model for other areas of mathematics. Perhaps this irrational confidence in the perfect harmony between geometric objects and their visible images was necessary to make possible the concept of mathematical models of processes in mechanics and physics which were to be so successful. It was not until the last third of the nineteenth century, with the study in depth of real numbers, that it was possible to take account of the gulf separating "geometric intuition" from the axioms supposedly providing their rational basis (see Chapter VI).

Criticisms of Euclidean constructions, which became very numerous, especially during the nineteenth century in the general movement towards greater "rigour" in mathematics (a topic to which we shall return in Chapter VI), do not, however, aim to correct the inferences made by Euclid in the course of his proofs, but rather the fact that they are not to a sufficient degree based on definitions and axioms which are made *explicit*. The general feeling was that if you could complete in a proper manner these foundations for reasoning, you could achieve an entirely satisfactory account. It was this task which was carried out by Pasch and Hilbert at the end of the nineteenth century when they enumerated systems of axioms made entirely explicit (twenty-three axioms in

Hilbert's case); by means of this all Euclid's theorems could at last be proved *without diagrams*.

Like Euclid, Hilbert begins with *undefined* notions, but he lists them *exhaustively*. There are three kinds of "primitive objects": points, straight lines and planes; and three "primitive relations": belonging (as a point does to a straight line or to a plane, for example), being "situated between" (as a point is in relation to two others, when all three belong to the same straight line), and being "congruent" (as with two segments or two angles[9]).

The question immediately arises as to how it is possible to reason correctly while refusing to define the things being discussed, and thus escaping an infinite regression in definitions. The answer is very simple: it is enough to force oneself *never* to state any proposition about geometrical objects and their relations which is not the logical consequence of the system of axioms governing them (these too being listed *exhaustively*). As Poincaré wrote [9], it can be said that these axioms constitute "disguised definitions" of the objects and relations with which they deal: these latter have in some sense disappeared, to be replaced by the sheaf of their "axiomatic" qualities.

Hilbert, following Pasch, pointed to a method of avoiding conclusions which might be suggested by geometric intuition, but are not derived from axioms: this would be to change the usual names of geometric objects and their relations. Hilbert proposed that we say "table", "chair" and "mug" instead of "point", "line" and "plane"[10]. For example, the first two axioms on Hilbert's list:

1) "Two distinct points belong to one and only one straight line",

2) "There are at least two distinct points belonging to the same line",

would become

1) "Two distinct tables belong to one and only one chair",

2) "There are at least two distinct tables belonging to the same chair".

It is clear that there would be no risk of involuntary error in statements such as these, devoid as they are of meaning in everyday language.

This may look like a joke; but in fact this dissociation of meaning and name embodies, for elementary geometry, the fundamental process which has

---

[9] Segments and angles are "derived" notions, that is, words whose definitions make use of certain axioms and primitive notions and relations, or of derived notions defined earlier: the segment AB is made up of points "between" A and B on the unique straight line Δ to which A and B belong; the half-line with origin A passing through B is made up of points of the segment AB and of points C of Δ such that B is "between" A and C; an angle $\{D_1, D_2\}$ is a pair of half-lines with the same origin.

[10] The possibility of arbitrarily choosing a word to define an object, that is, of abridging the statement of the properties which characterize it, was already noted by Plato (Letter VII, 343 b). The remark was taken up by d'Alembert, who declared in the *Encyclopedia* that there was nothing to stop us calling what is commonly known as a "circle" a "triangle".

freed mathematics from the bonds which tied it too closely to reality. It has made possible all the unlooked-for victories which have been won in the past century, and the amazing applications of these to physics. We shall return to this in detail in Chapter V.

5. Numbers and Magnitudes

While Greek mathematics was being elaborated into a "hypothetico-deductive" system[11] in the schools of philosophy, the necessities of daily life in Greek cities brought about, as in other civilizations, the existence of a class of professional "calculators". We know almost nothing of these "logisticians", as they were called[12], apart from their existence and the scorn which Plato expresses for them in the *Republic* (VII, 525), because they make calculations based on explicit *fractions*, while mathematicians must not, according to him, be concerned with anything other than natural numbers,

which are accessible only to the understanding, and which cannot be manipulated in any other way .

No doubt he meant by these words to refer to the admirable sequence of general theorems relating to *arbitrary* natural numbers which is to be found in Book VII of Euclid's *Elements* [13], where we find explained the elementary theory of the divisibility of natural numbers, of prime numbers, and of the decomposition of a natural number into prime factors.

This "logistic" tradition of numerical calculation comes to the surface only with Diophantus (about the fourth century A.D.). His techniques clearly extend those of the Babylonian tablets. They are designed to find one or several "unknown" numbers, which are solutions of a system of equations which we would write in the form of an equality between polynomials with specified coefficients. These equations could be of degree as high as six in the unknowns. For example (Book IV,19), the problem is to find three numbers such that each of the three numbers formed by taking the products of two of the three given numbers and increasing these by one is a perfect square. We would write this in the form of three equations

$$xy + 1 = u^2, \qquad yz + 1 = v^2, \qquad zx + 1 = w^2.$$

It is not necessary to examine here the methods of Diophantus, who only rarely appeals to general theories. What we want to emphasize is that he seeks a solution for each problem[14] in which the unknown quantities are natural

[11] The expression is Pieri's (1899).
[12] The Greek word which is translated as "logistics" means the practice of numerical operations.
[13] Euclid designates natural numbers by letters, and represents them by segments of lines.
[14] Even when the problem has several possible solutions, Diophantus almost always considers only one.

numbers or fractions of the form p/q (where p and q are natural numbers), or what we call *positive rational numbers*. Now in certain problems he comes up against *impossibilities* of two kinds, typified by these examples:

$$4 = 4x + 20 \quad \text{(Book V, 2)}$$

$$3x + 18 = 5x^2 \quad \text{(Book IV, 31)}.$$

For these cases Diophantus merely says that the problem is absurd. The second impossiblity results from the fact that 41 is not the square of a rational number, a generalization to natural numbers which are not squares of the discovery of the irrationality of $\sqrt{2}$, which we recounted in §2.

The importance of these impossibilities lies in the fact that they were overcome – the second in classical times, the first in the course of the Middle Ages – by the *creation of new mathematical objects*, far less close to obvious material images than those of the first Pythagoreans.

For Euclid and his predecessors, who did not want to calculate with fractions, these latter were replaced by the notion of the *ratio* between two lengths of segments or of polygonal lines, or of two areas of polygons, or of two volumes of solid bodies. The Greeks included lengths, areas and volumes under the general concept of "magnitude" (of which they form three different types), and Euclid's "common notions" can be seen as the beginning of the "Hilbertian" axiomatic treatment of this concept. In fact, Euclid *does not define* the notion of magnitude nor that of the two "primitive" relationships between magnitudes of the same type, that is, "being greater than" (which we would write as $A > B$), and "being the sum of two others" (which we would write as $C = A + B$). What he does is enumerate (but not exhaustively) some properties linking these notions; for example, his "common notion" number 4 would be written in our notation as:

$$\text{If } A > B, \quad \text{then } A + C > B + C.$$

A whole multiple pA of a magnitude A is obtained by adding the magnitude A to itself p times. Euclid then says that two magnitudes of the same type A and B are *commensurable* if there exists a third magnitude of the same type, C, such that $A = pC$ and $B = qC$ for two natural numbers p and q, and he says (Book X, 5), that A and B have the "same ratio" as the natural numbers p and q (which of course we write as $A/B = p/q$, in spite of Euclid).

This being the case, instead of saying, like Diophantus, that the diagonal of a square *has no* "ratio" to the side of a square, the Greek mathematicians of Plato's school declared that two magnitudes of the same type *always* have a ratio, even if they are incommensurable. They then succeeded, by means of a bold construction which was not fully understood until the nineteenth century, in defining for these general "ratios" concepts of inequality and of addition, beginning with these same familiar concepts used in dealing with ratios between natural numbers. We shall not describe here this remarkable invention (probably due to the greatest of all the mathematicians of this era,

Eudoxus of Cnidus), which is very close to our present-day conception (see Appendix 1).

Unfortunately, the system of calculation for these general "ratios" was not put into a usable form by the Greeks because of their conception of the *products* of ratios. In fact, when Euclid envisages the product of two ratios between *lengths* A/B and C/D (Book VI, 23), he shows that it is the ratio between two *areas* of rectangles $(A \times C)/(B \times D)$. Furthermore, Euclid has no definition of the sum of two arbitrary ratios, and, in the particular cases which he considers, the ratios have the same denominator (Book V, 24). In this system, then, the calculations of Diophantus (and even of his Babylonian predecessor cited in §1) are not practicable.

However, if we disregard philosophical preconceptions, it is not difficult to reconcile the two points of view, while remaining faithful to the Euclidean framework. After the resumption of scientific studies, at first in the Muslim world and then in the west, several mathematicians independently noticed this possibility: the poet-mathematician Omar Khayyam in the eleventh century, the Italian R. Bombelli in the sixteenth century, and finally Descartes, whose prestige brought about the definitive adoption of this reform. It consists in considering only the ratios between *lengths*, and reducing them all to the form OX/OU, where OU is a segment fixed once and for all and X is *any* point of the half-line Δ^+ extending OU. This is possible by a proposition of Euclid, asserting the existence of a "fourth proportional" (Book VI, 12). Instead of continuing to talk of "ratios", it is now simpler to say that we are considering the *points* X of Δ^+, and all we have to do is define the relation $X < Y$ between two points, and the operations $X + Y$ and XY, which must also define *points* and not other kinds of objects, as in Euclid.

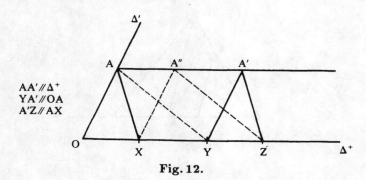

Fig. 12.

The definition of $X < Y$ is immediate: X must be "between" O and Y and different from Y. For $X + Y$ and XY, there are many ways of proceeding. The simplest seems to me to be a variant of that adopted by Hilbert, which requires only constructions of parallels, and is based only on the theorems of Euclid concerning congruent or similar triangles.

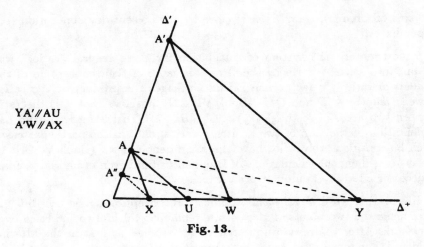

Fig. 13.

Consider a second half-line Δ' not extending Δ^+ and a point A on Δ'. The constructions of $Z = X + Y$ and $W = XY$ are shown in Figures 12 and 13. (They can of course be described without diagrams). The constructions of $Z' = Y + X$ and of $W' = YX$ are shown by dotted lines. These prove two of the fundamental properties of the calculus of the points of Δ^+:

$$Y + X = X + Y, \quad YX = XY.$$

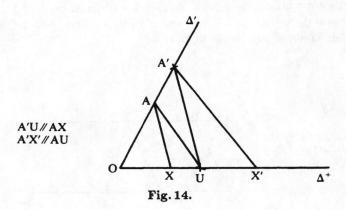

Fig. 14.

Figure 14 shows the existence, for all X of Δ^+, of the "inverse" point X' such that

$$XX' = X'X = U,$$

which we can also write as $1/X$ or X^{-1} (U being written also as 1, on account of the obvious equations $UX = XU = X$). Analogous reasoning makes it possible to prove all the properties listed in Appendix 2, except the last two.

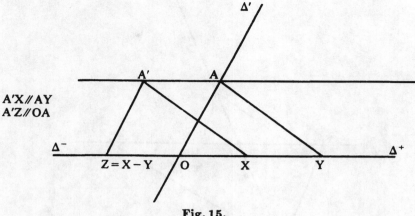

A'X//AY
A'Z//OA

Fig. 15.

It will be noted that in this presentation there was no need to make a distinction between commensurable and incommensurable ratios. An immediate extension of these constructions also makes it possible to do away with the first impossibility encountered by Diophantus. It was this problem which led Indian mathematicians in the fourth century A.D. to introduce "negative numbers". Their use spread only slowly, and was still meeting resistance at the end of the eighteenth century. In fact, all that is required is to *extend* the foregoing constructions to *all the points of the line* Δ which includes Δ^+, and not only to the points of Δ^+. The equation $Z + Y = X$ then always has a solution, even if $X < Y$ (Figure 15), and we write it as $X - Y$. In particular, $0 - X$ is written as $-X$, giving us $X + (-X) = 0$.

Nowadays we say that the points of Δ given by this construction are the *real numbers*, those of Δ^+ being *positive* and those of Δ^- being *negative*. It will be noted that these constructions "explain" at once the "rule of signs"

$$(-X)Y = -(XY),$$

and more particularly the fact that the product of two negative numbers is positive, something which often seems incomprehensible to non-mathematicians (Figure 16).

6. The Idea of Approximation

In the applications of mathematics, a real number Y is almost always replaced by a rational number X, which is an *approximating value* to it. That is to say, the difference $Y - X$ lies somewhere between $-\varepsilon$ and $+\varepsilon$, where ε is a positive number which is called the *approximation* to Y by X, or the *error* associated with Y; X is called an *approximating value to Y with margin ε*, where Y itself is taken as its "exact value". The Babylonians already knew

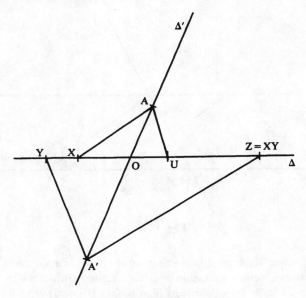

Fig. 16.

how to find rational approximating values for $\sqrt{2}$, and the Greeks invented general methods for obtaining approximating values for all square roots $\sqrt{a}$ (a being a natural number), within an *arbitrarily* fixed approximation.

The existence of such approximating values results from a general axiom explicitly formulated by Archimedes[15], often used by him, and customarily designated by his name: if Y and X are two positive numbers, there is always a natural number m such that $Y < mX$. If then p is the smallest of the natural numbers such that $Y < (p + 1)X$, we have $pX \leq Y < (p + 1)X$, whence pX approximates Y with margin X. An equivalent form of Archimedes' Axiom consists in stating that there is always a natural number m such that

$$\frac{1}{m}Y < X.$$

Thus there exists no number Z smaller than *all* other positive numbers (no "infinitely small" number). We may note in passing how far such an axiom is removed from sensory experience: it is impossible to make a material image of the operation of putting end to end segments of line equal to X until Y is passed unless X and Y are neither too big nor too small.

Since X can be subdivided into any number of equal parts (Euclid, Book VI, 10), once the approximate value pX is obtained we form the multiples

$$pX + m\frac{X}{10}$$

for $0 \leq m \leq 9$. If $q + 1$ is the smallest of those numbers for which

---

[15] It is used by Euclid in the proof of Book X, 1.

$$Y < pX + (q+1)\frac{X}{10},$$

then

$$pX + q\frac{X}{10}$$

is a value approximating Y with margin $X/10$. We continue with

$$pX + q\frac{X}{10} + r\frac{X}{100},$$

and so on.

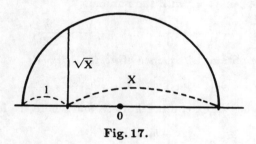

Fig. 17.

This procedure gives approximations to Y to within arbitrarily small margins, since for any number $Z > 0$ there exists a number n such that $X/10^n < Z$. It is assumed, however, that the procedure is being applied to an "exact" number X which is already *known*. But when $Z > 0$ is a known number, can the same be said of $X = \sqrt{Z}$? The construction of the "proportional mean" given by Euclid (Book VI, 13) (see Figure 17) answers this question by the intersection of a semi-circle and a line.

From the fifth century, however, the Greeks faced a similar problem with respect to what we call "cube roots". We shall see in §8 how they "constructed" them, and why their geometric conceptions prevented them from making similar "constructions" for "nth roots" in the case where $n \geq 4$.

After Diophantus these scruples scarcely arose any more. In the fourteenth century Oresme admitted implicitly that, for every natural number n and for every real number $X > 0$, there exists a number $Z > 0$ such that $Z^n = X$. We write this number as $X^{1/n}$. He even deduced from this the idea of the power $X^{p/q}$, with fractional exponent, and actually had the temerity to envisage powers with irrational exponents, such as $X^{\sqrt{2}}$. In the work of S. Stevin, in the sixteenth century, there appears an approximation procedure which generalizes the one which is described above, and which can at the same time be regarded as a proof of existence. He uses an equation, $P(t) = 0$, where $P(t)$ is a polynomial which we write as

$$P(t) = a_0 t^n + a_1 t^{n-1} + \ldots + a_n,$$

where $a_0 > 0$ and $a_n < 0$. From this we have $P(0) < 0$, and it is easy to show that there is a natural number $A > 0$ so large that $P(A) > 0$ (see Appendix 3). Stevin replaces the variable t in $P(t)$ successively by $1, 2, 3, \ldots$ and stops at the *last* natural number q such that $P(q) < 0$, so that $P(q+1) \geq 0$. He then replaces t by the numbers

$$q + \frac{1}{10}, \quad q + \frac{2}{10}, \quad \ldots, \quad q + \frac{9}{10},$$

and stops at the last natural number $q_1 \leq 9$ such that

$$P\left(q + \frac{q_1}{10}\right) < 0.$$

He continues in the same way with the numbers

$$q + \frac{q_1}{10} + \frac{1}{100}, \quad \ldots, \quad q + \frac{q_1}{10} + \frac{9}{100},$$

and so on. Thus he obtains a sequence of *nested intervals*

(1) $$[b_0, c_0], \quad [b_1, c_1], \quad \ldots, \quad [b_k, c_k], \quad \ldots$$

(Figure 18), which he can extend indefinitely, and where the length

$$c_k - b_k = 1/10^k.$$

We have in addition that $P(b_k) < 0$ and $P(c_k) \geq 0$.

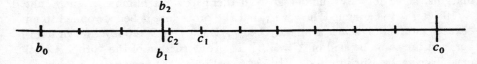

Fig. 18.

He then implicitly assumes that there is a number Z contained in *all* the intervals. Moreover, there cannot be two such numbers, $Z' < Z$, since then we would have $0 < Z - Z' < 1/10^k$ for *each* natural number k, contradicting Archimedes' Axiom. He next assumes that we have $P(Z) = 0$, which can be proved in an elementary way (see Appendix 3). This procedure gives him in addition a regular method of obtaining the *approximate values* $b_0, b_1, \ldots, b_k, \ldots$ of Z, with an arbitrarily small margin of error. The application of the method to the polynomial $P(t) = t^n - X$ gives a procedure for the approximation of $X^{1/n}$.

This procedure was to be used and generalized by the mathematicians of the succeeding centuries; but it was reserved for Cauchy to extract clearly from it the *Axiom of Nested Intervals*, affirming the existence of a real number in all the intervals of a sequence (1), in which each interval contains the next[16].

---

[16] We have in mind intervals which contain their endpoints.

In the axiomatic theory of real numbers, this axiom is the last of the axioms enumerated in Appendix 2, and it is without any doubt the most important, for it is the linchpin on which the whole of analysis turns.

7. The Evolution of Algebra

That it is possible to make calculations using unknown numbers as if they were known is not obvious, and it is often something which adolescents find difficult to understand. It is the real starting point for algebra, a threshold which we have already seen crossed by the Babylonian scribes of the second millennium B.C. But while it did not take more than a century for geometry to attain an essentially definitive form, it was not until thirteen centuries after Diophantus that algebra was to become what we know today.

The main obstacle which impeded its development was the way algebraists employed everyday language to describe calculations in algebra, using the fundamental operations, sum and product. This, for example, is the way the solution of a second-degree equation is described by al-Khwârizmî (ninth century), who was so famous in the Middle Ages that his name, a little corrupted, has become our word "algorithm":

Problem. *What must the square be which, when we add to its area* 10 *times the length of its side, is equal to* 39? *(in modern notation, $x^2 + 10x = 39$).*

Solution. *Divide by* 2 *the multiple of the side to be added, which gives* 5. *Multiply this number by itself, giving a product of* 25. *Add this number to* 39, *giving* 64. *Take the square root of this number, namely* 8 *and subtract from it half the multiple of the sides, leaving* 3. *This is the side of the required square, which is thus equal to* 9.

It is easy to imagine that, once the problem becomes a little bit complicated, the description of its solution, written in this way, must become just as obscure as legal jargon.

This difficulty led very gradually to the recognition of the need for a *shorthand* to make the sequence of operations easily comprehensible: here we have the problem of *notation*, which crops up again after every introduction of new objects, and which will probably never cease to torment mathematicians.

Diophantus was no doubt the first algebraist to use abbreviations. In a problem involving an unknown quantity, he uses special signs to denote its first six powers. The algebraists of the Middle Ages followed his example, writing

co or *c* (cosa)	the unknown x
ce (censo)	the square x^2
cu (cubo)	the cube x^3
cece (censo di censo)	the power x^4.

However, Diophantus and most of his successors did not have the notation for a second unknown quantity, when a problem involved one. At the end of the fifteenth century, Luca Pacioli wrote p and m for "plus" and "minus"[17], and used R or $R2$, $R3$, $R4$ or RR for the roots of orders 2,3 and 4. For example, he wrote

$$RV40\tilde{m}R320$$

for

$$\sqrt{40 - \sqrt{320}}$$

(the V stands for parentheses around the expression which follows. In the work of R. Bombelli we find L and ⌡ for parentheses). The signs + and − appeared only in about 1480, the sign = in 1557, and the signs for inequality, < and >, in 1631.

An essential step forward in notation is due to Viète: he used *letters* to designate not only unknowns but also *given* quantities, in a problem in which it is not necessary to give them numeric values. This already shortens considerably the description of calculations. For example, the formula which we would write as $bx^2 + dx = z$ is written by Viète as

B in A Quadratum, plus D plano in A, equari Z solido[18].

The notation for exponents (positive or negative) already used in the fifteenth century by N.Chuquet and M.Stifel, and then by Stevin (but not by Viéte) in the sixteenth, became current usage with the work of Descartes. Finally there remained the introduction of *indices* for a sequence of numbers when we do not want to specify how many terms there are. This idea occurs in the work of Newton and of Leibniz, but up to the end of the nineteenth century many mathematicians preferred to write

$$a, b, c, \ldots, l,$$

rather than the modern notation

$$a_0, a_1, \ldots, a_n,$$

which enables us to calculate with $n + 1$ terms without inconvenience. Thus it is only at the end of this long evolutionary process that it became possible to write a "general polynomial"

$$a_0 x^n + a_1 x^{n-1} + \ldots + a_n = \sum_{j=0}^{n} a_j x^{n-j},$$

and this was to make it possible to express and to prove general theorems in algebra.

---

[17] For "piu" and "meno", respectively.
[18] Viète, faithful to the Platonic tradition, adds together only measurements of the same kind: thus B and A are lengths, D an area, Z a volume.

8. The Method of Coordinates

It is well known that we owe to Descartes (and simultaneously to Fermat) one of the three most important innovations in mathematics of the seventeenth century, the introduction into geometry of the *method of coordinates* (also known as the "application of algebra to geometry", and later as "analytic geometry"). They were not, however, without predecessors, since the first examples of plane curves defined by equations go back to the fourth century B.C. We have seen (§2) that the construction of a square twice the area of another is easy: the analogous problem of the "duplication of the cube" no doubt presented itself quite early on, but it seemed to the Greek mathematicians to be much more difficult. The duplication of the square amounted to the construction of a length x which is the "proportional mean" between two lengths a and $2a$, that is, in such a way that

$$\frac{2a}{x} = \frac{x}{a}$$

since, by the definition of the product of two ratios (§5):

$$\frac{x^2}{a^2} = \frac{2a \times x}{x \times a} = 2\,.$$

It was this way of putting the problem which led Hippocrates of Chio (fifth century) to reduce the duplication of the cube to the construction of *two* "proportional means" between a and $2a$, that is to say, two lengths x and y, satisfying the equations

$$(2) \qquad \frac{2a}{x} = \frac{x}{y} = \frac{y}{a}\,.$$

In fact, for the Greeks, the product of the three lengths A, B and C is the *volume* of the parallelopiped of sides A, B, C, and thus the relations (2) entail (Euclid, Book VIII, 12, for the commensurable ratios)

$$\frac{y^3}{a^3} = \frac{2a \times x \times y}{x \times y \times a} = 2\,.$$

This construction did not seem any easier to carry out than the original problem. But Menaechmus, a pupil of Eudoxus, observed that (2) could also be written in the form of two simultaneous equations

$$(3) \qquad x^2 = 2ay$$

$$(4) \qquad xy = 2a^2.$$

He then had the idea of taking two perpendicular half-lines, OX and OY in a plane, and considering *separately* the relations (3) and (4) between a segment $x = OP$ on OX and a segment $y = OQ$ on OY. But for *each*

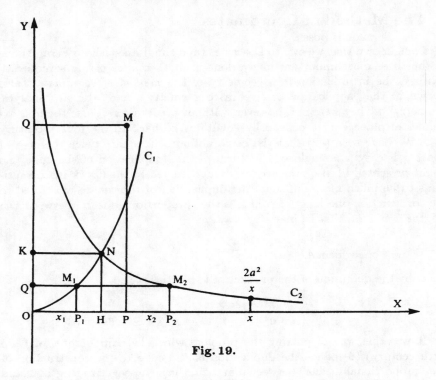

Fig. 19.

point Q on OY with $OQ = y$, we have, on the line parallel to OX drawn through Q (Figure 19), a point M_1 such that $x_1 = OP_1 = QM_1$ satisfies $x_1^2 = 2ay$ (a "proportional mean" between $2a$ and y), and a point M_2 such that $x_2 = OP_2 = QM_2$ satisfies $x_2 y = 2a^2$ (a "fourth proportional" between $2a$, y, and a).

When Q varies on OY, M_1 and M_2 describe two curves C_1 and C_2. The same Menaechmus discovered a little later that the curves could be obtained as the plane sections of a cone of revolution. After Euclid and Archimedes, these plane sections, known as *conics*, were studied in depth by Apollonius (third century B.C.), who named C_1 a *parabola* and C_2 a *hyperbola*.

If N is the point of intersection of these two curves, the segments $OH = NK = x$ and $OK = NH = y$ give the solution of (2). Anecdotal history records that Plato did not like this solution because it used curves other than the straight line and the circle; but this type of "construction" by the intersection of curves became none the less popular among Euclid's successors (even if it did imply properties that were "visible" but not proved), and it was still highly prized by Descartes and Newton[19].

---

[19] One could define in the same way three, four, ... "proportional means" between $2a$ and a, but the Greeks would not have been able to make "constructions" of the roots of order > 3, since the product of more than three lengths would have meant nothing to them.

The importance of the method used by Descartes and Fermat lies in the fact that it made it possible to *translate any problem in plane geometry into an equivalent problem in algebra*. This is because the objects and relations of Hilbert's axiomatic system can be *interpreted* as objects and relations of the *theory of real numbers*, in accordance with the following "dictionary":

point M	pair of numbers (x, y); we write $M = (x, y)$
straight line	equation $ax + by + c = 0$ with $a^2 + b^2 \neq 0$
point belonging to ...	pair satisfying ...
point between $A = (x_1, y_1)$ and $B = (x_2, y_2)$	pair $(tx_1 + (1-t)x_2, \ ty_1 + (1-t)y_2)$ with $0 < t < 1$
length of segment AB	$((x_1 - x_2)^2 + (y_1 - y_2)^2)^{1/2}$
circle	equation $x^2 + y^2 + ax + by + c = 0$ with $4c < a^2 + b^2$

Finally, the formula for the angle α between the lines AB and AC is

$$\cos \alpha = \frac{(x_1 - x_0)(x_2 - x_0) + (y_1 - y_0)(y_2 - y_0)}{((x_1 - x_0)^2 + (y_1 - y_0)^2)^{1/2} ((x_2 - x_0)^2 + (y_2 - y_0)^2)^{1/2}},$$

where $A = (x_0, y_0)$, $B = (x_1, y_1)$, and $C = (x_2, y_2)$.

"Geometry in space" can be "translated" in an analogous way by substituting for a point a combination of three numbers (x, y, z), for a plane an equation $ax + by + cz + d = 0$ with

$$a^2 + b^2 + c^2 \neq 0, \quad \text{etc.}$$

In the nineteenth century, when the foundations of mathematics came to be given more depth, it would be said that elementary geometry has a *model* in the theory of real numbers. This is the first of a series of "bridges" linking two dissimilar parts of mathematics. We shall come across some of these in Chapter VI, clearer and clearer evidence of the fundamental *unity* of mathematical ideas.

Furthermore, the fact that the lines, circles and conics in a plane are defined by equations of the form $P(x, y) = 0$, where P is a polynomial of the first or second degree with real coefficients[20], naturally led mathematicians to study curves defined by the same kind of equation but without restriction as to degree. This was the beginning of a new branch of mathematics, *algebraic geometry*, which added enormously to the list of curves already known to the

---

[20] A polynomial in x and y is the sum of *monomials* $a_{jk}x^j y^k$. Its *degree* is the largest of the numbers $j + k$.

Greeks, and which is nowadays still one of the most active branches, after three hundred years of research and a mass of results.

The method of coordinates is also at the base of two other great advances made in the seventeenth century: the introduction of the idea of *function* and the infinitesimal calculus. It is often said that the mathematical concepts of the Greeks were fundamentally *static*, and to this is opposed the idea of variation which dominates modern scientific thought. It is certainly true that Euclid's *Elements* are centred on the study of figures whose positions and magnitudes are *fixed*. Nevertheless, since the beginnings of Greek thought, attempts to understand movement and change in form or in nature continually occupied the minds of the philosophers, and the notions of *uniform* movement – rectilinear or circular — had been clearly set in view ever since it had been known how to measure time. We know that, in their astronomical systems, it was by combinations of these movements that the Greeks tried to give an account of the trajectories of the planets. And although the concept of time is not included in Greek geometry[21], at least two plane curves, the quadratrix of Hippias and the spiral of Archimedes, were defined by combinations of uniform movements[22].

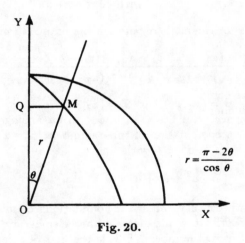

$$r = \frac{\pi - 2\theta}{\cos \theta}$$

Fig. 20.

It seems to have been principally for the study of rectilinear movements which are not necessarily uniform – notably of falling objects, a subject which figured prominently in the philosophical schools of the Middle Ages – that Oresme in the fourteenth century, probably for the first time, had the idea of

---

[21] Perhaps we should consider the famous paradoxes of Zeno of Elea as resulting from unsuccessful attempts to bring time and movement into a "hypothetico-deductive" system.

[22] The quadratrix of Hippias is described by a point M such that OM turns around O in a uniform movement, and the projection Q of M on OY moves uniformly on OY (Figure 20). The spiral of Archimedes is described by a point M such that D turns at constant speed around O and M moves uniformly along D (Figure 21).

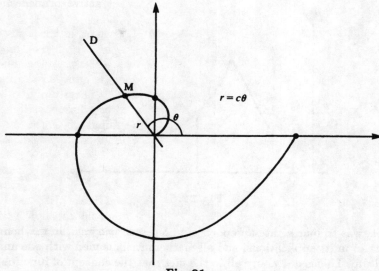

Fig. 21.

representing a variation in magnitude which occurs over time by a *graph*, in which the abscissa of a point on the graph represents time and the ordinate represents the value of the varying magnitude at that time (Figure 22); the points thus obtained constituting the graph. Oresme also considered that, instead of time, one could use as the abscissa any "quality" which could be represented by a number[23]. In our day the procedure has become ubiquitous, and is often misused.

In the seventeenth century this idea was combined with the method of coordinates, to make familiar the idea of a number y "dependent" on a number x which varies over an interval I. At the end of the century it would be said that y is a *function* of x: its graph is thus a curve intersected at only one point by each of the lines passing through a point of I and parallel to OY. *Conversely*, however, each curve having this property defines a function: for example, the semi-circle of centre O and radius 1 situated above OX (Figure 17) defines in the interval $-1 \leq x \leq 1$ the function

$$y = \sqrt{1 - x^2}.$$

It is this correspondence which in the nineteenth century was to make it possible to define a general concept of function as a mathematical object in the platonic sense (see Chapter V, §3, B). But until then there was scant concern for theoretical foundations. The concept of function, however "intuitive" it

---

[23] This is not necessarily a "magnitude" as the Greeks understood it. An example is temperature, measured in degrees centigrade: the concept of one temperature being smaller than another has a physical meaning, but not that of a temperature of which the measure would be the "sum" of two others.

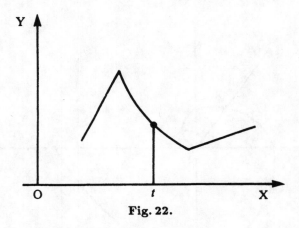

Fig. 22.

might be, was to inaugurate an epoch of unforeseen progress, in mathematics no less than in its applications, and all mathematicians vied with one another to take it up. This is due above all to the fact that the concept of function is at the base of the third invention of the seventeenth century – perhaps the most important in the whole history of mathematics – the infinitesimal calculus.

9. The Concept of Limit and the Infinitesimal Calculus

The first example of the determination of a number by the process of "nested intervals" (§6) which has come down to us is the approximation to the area of a circle, given by Euclid (*Elements*, Book XII, 2). He inscribed in a circle a square P_1, then, taking one after another the mid-points of the arcs subtended by the sides of the polygons, he obtained the regular polygons $P_2, P_3, \ldots, P_n, \ldots$ with $8, 16, \ldots, 2^{n+1}, \ldots$ sides. At the same time, he considered the circumscribed polygons, $Q_1, Q_2, \ldots, Q_n, \ldots$, whose sides are tangents to the circle at the points of the inscribed polygons (Figure 23). Let p_n be the area of the polygon P_n and q_n the area of the polygon Q_n. Then these numbers are the end points of a sequence of nested intervals

$$[p_1, q_1], \quad [p_2, q_2], \quad \ldots, \quad [p_n, q_n], \quad \ldots,$$

and Euclid shows by a beautiful geometric argument that

$$q_n - p_n \leq \frac{1}{2}(q_{n-1} - p_{n-1}).$$

He is thus able to assert that the unique number contained in all these intervals measures "the area of the circle" (see Appendix 4).

This construction is simply an example of what was called the "method of exhaustion", probably invented by Eudoxus. In any case it was he who applied it to prove that the volume of a cone of revolution is a third of the volume

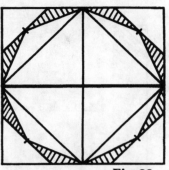

"The hatched part has area $q_2 - p_2$"

Fig. 23.

of a cylinder of the same base and the same height (Euclid, Book XII, 10). In the hands of Archimedes it produced a whole harvest of new results: the area of a segment of a parabola, the area and volume of a sphere, the volume of a segment of a paraboloid of revolution, etc.

In the seventeenth century, the use of coordinates made it possible to generalize the method of exhaustion. For example, if in an interval

$$I \; : a \leq x \leq b,$$

$y = f(x)$ is a positive increasing function, we can evaluate by this method the area S contained between its graph and the axis OX: we divide I successively into $2, 4, 8, \ldots, 2^n, \ldots$ equal parts, and, for each partial interval $\left[t_k, t_{k+1}\right]$, we consider the area

$$\frac{(b-a)}{2^n} \cdot f(t_k)$$

of the rectangle with base this interval which is situated "under" the curve, and the area

$$\frac{(b-a)}{2^n} \cdot f(t_{k+1})$$

of the rectangle with the same base and "above" the curve. If S'_n is the sum of the areas of the first rectangles, and S''_n is the sum of the areas of the second, then we have $S'_n \leq S \leq S''_n$ and

$$S''_n - S'_n = \frac{(b-a)}{2^n}\left(f(b) - f(a)\right) \qquad \text{(Figure 24)}.$$

The number S is thus defined as the unique number contained in all the nested intervals $\left[S'_n, S''_n\right]$. This is the way in which Fermat proceeded from 1636 onwards with respect to the curve $y = x^m$ (where m is a natural number with $m > 1$), by a simple algebraic formula which gives S'_n and S''_n explicitly, and thus he obtained the formula

$$S = \frac{b^{m+1} - a^{m+1}}{m+1}$$

(see Appendix 4).

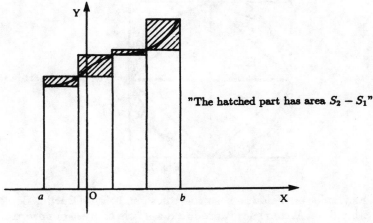

"The hatched part has area $S_2 - S_1$"

Fig. 24.

Later, by a better adapted choice of the division of I into smaller and smaller intervals, he extended these results to the case where $m = p/q$ is any fraction (other than -1).

This method (with minor improvements making possible smaller "nested intervals") is still one of those which form the basis for the calculation of areas by computers. It would not perhaps have raised much interest if it were not for the fact that the same mathematicians were at the same time attacking a problem which looks very different, one merely touched on in classical times, that of the determining the *tangents* to a plane curve.

For the Greeks a tangent to a curve C at one of its points M is a line D passing through M such that the points of C *near to* M are all on the same side of D (Figure 25): D "touches" the curve at the point M (in English the word should be "touching" rather than "tangent"). Euclid shows that the tangent at a point M of a circle with centre O is perpendicular to OM (Book III, 16), and the description of conics as plane sections of cones with circular bases made it possible to find tangents to these curves. Apart from conics, however, only one curve is known whose tangents were determined in classical times, that is Archimedes' spiral (§8), and we have no idea how he guessed the construction of these tangents[24].

It was Fermat again who, in 1636, used the method of coordinates to attack the problem in a systematic way for the curves $y = x^m$ (where m is a natural number ≥ 2). The question is whether or not a line D, given by the equation $Y = aX + b$, passing through the point $M = (x, y)$ with $y = x^m$, "touches" this curve at this point. For a point P of OX, with abscissa $x + h$ with $|h|$ *sufficiently small*, it is thus necessary to know the *sign* of the directed segment $\overline{RS}$ (Figure 26). Since D passes through M, we have $x^m = ax + b$, and

---

[24] He carries out the construction of the line without any explanation, then proves by an "apagogical" argument (Appendix 1) that this line cannot cross the curve at the point of contact.

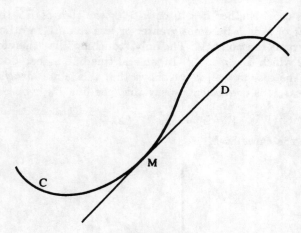

Fig. 25.

$$\overline{RS} = (x+h)^m - (a(x+h)+b)$$
$$= mhx^{m-1} + h^2 Q(x,h) - ah$$
$$= h(mx^{m-1} - a + hQ(x,h))$$

by the binomial formula, where $Q(x,h)$ is a polynomial in x and h, and there is a number A such that $|Q(x,h)| \leq A$ for $|h| \leq 1$ (Appendix 3).

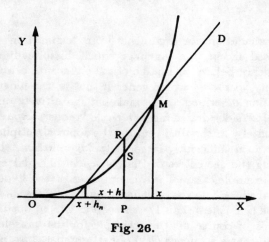

Fig. 26.

If the number $\alpha = mx^{m-1} - a$ is not zero, the parenthesis in the last formula has the *same sign as α* as soon as

$$|h| < 1 \quad \text{and} \quad |h| < |\alpha|/A,$$

and this is the case whether $h < 0$ or $h > 0$; the sign of $\overline{RS}$ thus *changes* along with that of h as h becomes greater or less than 0. In other words, D *crosses* the curve at the point M. The only "touching" line therefore must be the line D_0 for which $a = mx^{m-1}$. In general this line indeed does not cross the curve, but the case $m = 3$, $x = 0$ shows that this is not always so (Figure 27). Nevertheless, it is convenient to say that the line D_0, whose equation is therefore

$$Y - x^m = mx^{m-1}(X - x)$$

is a tangent to the curve at M.

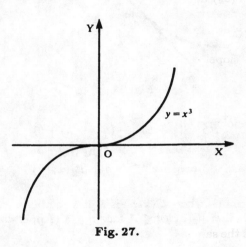

$y = x^3$

Fig. 27.

Neither this calculation nor arguments from "exhaustion" could be satisfactorily expressed except in a language which, historically speaking, was not clearly in evidence before the middle of the eighteenth century and which took another eighty years to become general: that is, the language of *limits*. Euclid's construction described above makes it "intuitively obvious" that a regular polygon of 2^n sides drawn inside a circle "tends" towards the circle as n increases indefinitely. In the fifth century the sophist Antiphon went so far as to say that for an n sufficiently large the polygon *became* the circle itself, deeply scandalizing the geometers of Plato's school. The latter had at their disposal another example, that of the asymptotes of the hyperbola (Figure 19), which may have been known to Menæchmus: the distance from the point $(x, 1/x)$ to the line OX is less than $1/n$ as soon as $x > n$, but *is never zero*[25], something which is of course only conceivable for the "absolute figures" of Plato. We now say that a sequence (h_n) of real numbers has *limit* 0, or *tends to* 0, when, for *every* number $\alpha > 0$, all of the h_n belong to the interval $-\alpha < x < \alpha$ *for all n from a certain number* n_0 *onwards* (where of course n_0

---

[25] This statement is difficult for a non-mathematician to grasp. Montaigne cites it as an oddity of "learned men". This kind of conceptual difficulty can help us understand why it took so long to reach the general concept of limits.

depends on α). More generally, (h_n) *tends to* an arbitrary real number a, or has *limit* a if the sequence $(h_n - a)$ tends to 0. We then write

$$\lim_{n \to \infty} h_n = a.$$

Using this language, the Axiom of Nested Intervals (§6) is expressed by saying that, if

$$\lim_{k \to \infty} (c_k - b_k) = 0,$$

then the sequences (b_k) and (c_k) have *the same limit*, the unique number Z contained in all the intervals.

Returning to Fermat's calculation for determining the tangent to a curve $y = x^m$, we can say that the "slope" $a = mx^{m-1}$ of the tangent at point M is the *limit* of the "slope"

$$(5) \qquad \frac{1}{h_n} \left((x + h_n)^m - x^m \right)$$

of a "secant", a line joining M to a point on the curve with abscissa $x + h_n$ (and which is defined only if $h_n \neq 0$), for *each sequence* (h_n) *of non-zero numbers tending to* 0. In this form the definition can be extended to *each* curve with the equation $y = f(x)$, on condition that the limit of the sequence

$$(6) \qquad \frac{1}{h_n} \left(f(x + h_n) - f(x) \right)$$

exists and is always the same, for all sequences (h_n) of non-zero numbers which tend to 0.

It was soon realized, in the seventeenth century, that Fermat's method could be applied to many other curves known at the time, for example, to $x^3 y^2 = 1$, $y = \sin x$, $y = \tan x$, etc. For modern kinetics and dynamics, which were coming into being at the same time through the work of Galileo and Kepler, this same "approach to a limit" provided a definition of "instantaneous speed" – a concept known intuitively up till then, but not precisely formulated – and above all the concept of acceleration, which was to become the cornerstone of Newtonian mechanics.

It must not be thought that this short résumé reflects the way mathematical thought progressed in creating the infinitesimal calculus during the seventeenth century: the way was slow, tortuous and confused. In expression (5), the numerator and the denominator *both* become 0 when we take $h_n = 0$, and the expression 0/0 has no meaning. Mathematicians tried to find a way out by speaking, as Leibniz did, of the "infinitely small"[26], or, as Newton did,

---

[26] Modern axiomatic methods enable us to define rigorously a set $\mathbf{R}^*$ of "non-standard numbers" which contains the set $\mathbf{R}$ of real numbers (also called "standard numbers") but in which there are also "infinitely small numbers". By using these numbers Leibniz's reasoning can be entirely justified: it constitutes what is known as "non-standard analysis", which often provides simpler proofs of theorems in standard analysis, and which can even lead to the discovery of new properties.

of "ultimate values of vanishing quantities", but this is only to cover up with words the imprecision of the ideas. Nevertheless, by the end of the century, increasing familiarity with this type of reasoning and its obvious fruitfulness led to its being codified through the introduction of a common notation and of algorithms which made it possible to manipulate them readily. The limit of expression (6) defines, as x varies, a new function which Newton called the "fluxion" of the function $y = f(x)$ (which he termed the "fluent") and he gave the notation $\dot{y}$ to this function, which Leibniz for his part would write as dy/dx. Later the limit of (6) would be written $f'(x)$, and it would be said that this is the *derivative* at the point x of the function $y = f(x)$. As for the area S of Figure 24, the term *integral of the function $y = f(x)$ between a and b*, and the notation

$$\int_a^b f(t)\,dt\,,$$

which was to be used from the time of Fourier, descend directly from the terms used by Leibniz. Moreover, the terminology and the notation were extended to functions which are not necessarily positive in the interval I by taking as negative those parts of the area below the axis OX.

The great leap forward of the infinitesimal calculus was made possible by the key discovery of the *link* between derivatives and integrals. This was made independently by many mathematicians in about 1660, in a more or less confused way. When the point x varies over the interval I, the formula

$$F(x) = \int_a^x f(t)\,dt$$

defines a *function* of x; and, assuming the hypotheses set out above, the function $y = F(x)$ has at every point x *a derivative equal to $f(x)$* (we say that $F(x)$ *is a primitive* of the function $f(x)$). The proof is easy. For $h > 0$, we have (Figure 28)

$$h \cdot f(x) \le \int_x^{x+h} f(t)\,dt \le h \cdot f(x+h),$$

which can also be written

$$0 \le \frac{1}{h}(F(x+h) - F(x)) - f(x) \le f(x+h) - f(x),$$

and by hypothesis $f(x+h) - f(x)$ is arbitrarily small with h.[27] The reasoning is the same for $h < 0$.

(Once one primitive $F(x)$ of a function $f(x)$ is known, all the others are functions $F(x) + C$, where C is an arbitrary constant. Leibniz writes

$$\int f(x)\,dx$$

---

[27] Until about 1800, all functions were implicitly assumed to be continuous (in our sense) (Appendix 3).

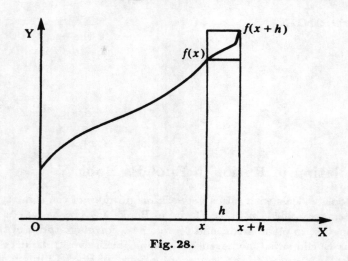

Fig. 28.

for an *arbitrary* primitive of f.)

This fundamental result makes it possible to pass from properties of the derivative to those of the integral, and vice-versa. In Appendix 5 we shall give some examples of applications of algorithms obtained in this way, which make the calculus a tool of extraordinary flexibililty.

In conclusion, we observe that all the geometrical ideas described above, like all the other geometrical applications of the calculus up until the middle of the nineteenth century, are based on *unformulated* hypotheses about the length of a curve, the area of a surface, and the volume of a solid. They are suggested by "geometric intuition", that is in fact by simplified diagrams. In order to transform these concepts into "mathematical objects" in Plato's sense, difficult problems would need to be surmounted in the nineteenth and twentieth centuries.

III Appendix

1. Calculation of Ratios in Euclid's Book V

Euclid's Book V deals with ratios between magnitudes of the same type, this type not being specified: in Book VI and in Books X – XIII he makes applications to lengths, to plane areas, and to volumes. For magnitudes of the same type (or, as Euclid says, "homogeneous among themselves"), he uses the two relationships introduced in the "common notions" of Book I : in our notation $A < B$ and $C = A + B$. In Definition 5 in Book V, he accepts in addition that two magnitudes of the same type satisfy "Archimedes' Axiom" (§5). For two magnitudes of the same type, A and B, Euclid introduced a new object, their *ratio*, which he did not define (Definition 3 is a "pseudo-definition", unusable, like those of the point or the line (§4)). We write A/B for brevity's sake, but Euclid has no notation and speaks always of "the ratio of A to B". Next, taking these ratios, he *defined* the relations between them, $A/B = C/D$ (Definition 6) and $A/B < C/D$ (Definition 8), using the relationships $A = B$ and $A < B$ among magnitudes of the same type. Until the nineteenth century these definitions seemed to most mathematicians unnecessarily complicated, and even incomprehensible. I should like to explain here that this seeming complexity is *necessitated* by the existence of incommensurable ratios. It illustrates the depth and subtlety of Greek thought.

For ratios between integers, the relationship of equality, $m/n = p/q$, is equivalent to $mq = np$, and that of inequality, $m/n < p/q$ to $mq < np$. But this relationship $m/n < p/q$ can also be expressed in the following manner: there exists a ratio r/s between integers such that

$$(1) \qquad\qquad m/n < r/s < p/q$$

(if we want to find such a ratio when we know that $m/n < p/q$, we need only take $r = m + p$ and $s = n + q$). A third way would be to say: there exist two integers r and s such that $rq < sp$, but $rn > sm$.

Now if the ratios $A/B = m/n$ and $C/D = p/q$ were commensurable, we could, in virtue of what has just been said, express the relationship $A/B < C/D$ in the form:

There exist two whole numbers r and s such that $rD < sC$ but $rB > sA$.

In this formulation, however, the fact that the ratios A/B and C/D are commensurable *is irrelevant*. This is *exactly*, Definition 8 of Book V for *any* ratios, commensurable or not.

Now how are we to define the relationship of equality $A/B = C/D$? Simply by saying that we have *neither* the inequality $A/B < C/D$ *nor* the inequality $C/D < A/B$, from which comes Definition 6 of Book V for $A/B = C/D$:

For all pairs r,s of integers such that $rD < sC$ (respectively, $rD = sC$, $rD > sC$), we have $rB < sA$ (respectively, $rB = sA$, $rB > sA$).

When therefore the Greek mathematicians, beginning with Euclid, wished to prove an equality $A/B = C/D$ of two ratios between magnitudes, they employed what is known as an *apagogical* argument, which consists in *two* proofs by contradiction: first we prove that the existence of two integers such that $rD < sC$ and $rB > sA$ is absurd, and then that the same absurdity arises if we invert the roles of A/B and C/D. It is in this way, for example, that Euclid, using the process of "exhaustion" of §9, proves that the ratio of the areas of two circles is the *square* of the ratio of their radii (Book XII, 2), and (by Eudoxus' method of exhaustion) that the ratio of the volume of a cone to that of a cylinder of the same base and the same height is equal to 1/3 (Book XII, 10).

2. The Axiomatic Theory of Real Numbers

We shall see in Chapter VI why the identification of real numbers with points on a line, originating with Descartes (§4), eventually gave rise to misgivings on the part of nineteenth-century mathematicians, preoccupied as they were with "rigour", and shall see how they preferred "arithmetical definitions" of real numbers. On the other hand, once the axiomatic method implemented by Hilbert in geometry (§4) was properly understood, it became apparent that it could be applied to many other areas of mathematics, and it was in this way that, about 1930, algebra, topology and the theory of topological vector spaces were set out (see Chapter V, §5, A). It has however only recently been realized that the same method can be applied to the real numbers, and that, in so far as we do not concern ourselves with questions of non-contradiction, this makes it possible to build a foundation for the whole of analysis which dispenses with the complications of "arithmetical definitions" (Chapter VI, Appendix 2).

The theory is simpler than the axiomatic system of Euclidean geometry. There is only one kind of "primitive object", known as "real number", and between the real numbers there are only three "primitive relationships" :

1) The relationship $x \leq y$ (also written $y \geq x$) between two real numbers ("order").

2) The relationship $z = x + y$ between three real numbers ("addition").

3) The relationship $z = xy$ between three real numbers ("multiplication").

The theory requires seventeen axioms:

R 1) There are two distinct real numbers.

R 2) If x and y are two real numbers, then either $x \leq y$ or $y \leq x$.

R 3) If at the same time $x \leq y$ and $y \leq x$, then $x = y$, and conversely.

R 4) If $x \leq y$ and $y \leq z$, then $x \leq z$.

R 5) $(x + y) + z = x + (y + z)$.

R 6) $y + x = x + y$.

R 7) There exists a real number α such that $\alpha + x = x$ for each real number x.

R 8) For each real number x, there exists a real number x' such that $x + x' = \alpha$.

R 9) If $x \leq y$, then we have $x + z \leq y + z$ for each real number z.

R 10) $x(yz) = (xy)z$.

R 11) $yx = xy$.

R 12) There exists a real number ε such that $\varepsilon x = x$ for each real number x.

R 13) For each real number $x \neq \alpha$, there exists a real number x'' such that $xx'' = \varepsilon$.

R 14) If $\alpha \leq x$ and $\alpha \leq y$, then $\alpha \leq xy$.

R 15) $x(y + z) = xy + xz$.

For each real number x, we define by induction on n the real number $n \cdot x$ by

$$1 \cdot x = x, \quad (n + 1) \cdot x = n \cdot x + x.$$

R 16) If $\alpha \leq x$ and $\alpha \neq x$, then, for each real number y, there exists an integer n such that $y \leq n \cdot x$ ("Archimedes' Axiom").

R 17) If $(b_k)_{k \geq 0}$ and $(c_k)_{k \geq 0}$ are two infinite sequences of real numbers such that, for every $k \geq 0$,

$$b_k \leq b_{k+1} \leq c_{k+1} \leq c_k,$$

then there exists a real number x such that

$$b_k \leq x \leq c_k$$

for each k ("Axiom of Nested Intervals").

As an example of the way in which an axiomatic theory is developed, here are the proofs of the first properties deduced from the axioms:

1) The number α which satisfies R 7) is *unique*. For, if α' is a second real number such that $\alpha' + x = x$ for each x, we have, by R 6),

$$\alpha = \alpha' + \alpha = \alpha + \alpha' = \alpha'.$$

From now on we write 0 for the unique number α which satisfies R 7).

2) For each real number x, the real number x' which satisfies R 8) is *unique*. For suppose that $x + x' = 0$ and $x + y = 0$. Then, by R 5) and R 6),

$$x' = x' + 0 = x' + (x + y) = (x' + x) + y$$
$$= (x + x') + y = 0 + y = y.$$

From now on we write $-x$ for the unique number x' which satisfies R 8). It follows that $-(-x) = x$.

3) The relationship $x + z = y + z$ implies that $x = y$. For

$$x = x + 0 = x + (z + (-z)) = (x + z) + (-z)$$
$$= (y + z) + (-z) = y + (z + (-z)) = y + 0 = y.$$

4) For each real number x, $0x = 0$. For, by R 15),

$$0x + yx = (0 + y)x = yx = 0 + yx,$$

from which, by 3), $0x = 0$.

Since there exists an $x \neq 0$ by R 1), it follows from R 12) and from 4) that $\varepsilon \neq 0$. The condition $x \neq 0$ in R 13) is therefore necessary.

5) If $xy = 0$, *at least one* of the two numbers x and y is 0 (we say that there is no real number "which is a zero-divisor"). For, if $x \neq 0$, then we have, by R 10), R 11), R 12), and R 13), that

$$x''(xy) = (x''x)y = \varepsilon y = y \quad \text{and} \quad x''(xy) = x''0 = 0x'' = 0.$$

The relation "$x \leq y$ and $x \neq y$" is written as $x < y$ or as $y > x$. Then, if $x < y$ and $y \leq z$, it follows that $x < z$, because $x \leq z$ by R 4), and if $x = z$, it would follow that $y \leq x$, and hence, by R 3), that $x = y$, contradicting the hypothesis that $x \neq y$. It can be proved in the same way that if $x \leq y$ and $y < z$, then $x < z$.

6) If $x > 0$, then $-x < 0$. For it is not possible to have $-x = 0$, for else it would follow that $x = -(-x) = 0$. Neither is it possible to have $-x > 0$, for else it would follow by R 9) that $0 = x + (-x) \geq x + 0 = x > 0$, which is absurd.

7) We have $(-x)y = -(xy)$ and $(-x)(-y) = xy$ (the rule of signs). For, by 2), 4), and R 15), we have

$$0 = 0y = (x + (-x))y = xy + (-x)y.$$

Hence $(-x)(-y) = -(x(-y)) = -(-(xy)) = xy$.

8) For each $x \neq 0$, we have $x^2 > 0$. For, if $x^2 = 0$, then $x = 0$ by 5). If $x > 0$, it follows that $x^2 \geq 0$ by R 14), and it is not possible that $x^2 = 0$. Hence, if $x < 0$, it follows that $-x > 0$ and $x^2 = (-x)^2$ by 7).

9) The number ε which satisfies R 12) is unique.

10) For each $x \neq 0$, the number x'' which satisfies R 13) is unique.

The proofs follow the same course as those of 1) and 2), except that multiplication replaces addition. We write 1 for the unique number ε which satisfies R 12) and x^{-1} or $1/x$ for the unique number x'' which satisfies R 13). It is clear that $1 = 1^2 > 0$, and hence $-1 < 0$. We have $(-1)x = -x$, since

$$x + (-1)x = 1x + (-1)x = (1 + (-1))x = 0x = 0.$$

The other standard properties of the real numbers are proved in the same manner.

For each natural number $n > 0$, n is identified with the real number $n \cdot 1$. The set Z of real numbers 0, $n \cdot 1$ and $n \cdot (-1) = -(n \cdot 1)$ for all the natural numbers n, is known as the set of *integers*.

3. Approximation of the Real Roots of a Polynomial

The method of Stevin described in §6 is based on the following *lemma* (which he does not prove).

If, for a polynomial $P(x)$, we have $P(x_0) > 0$ for a number x_0, then there exists an *interval* $x_0 - \alpha < x < x_0 + \alpha$ (where $\alpha > 0$) such that $P(x) > 0$ at *all* the points of this interval.

Let us suppose that this lemma has been proved. Then, with the notation of §6, it can be proved by an "apagogical" argument (Appendix 1) that $P(Z) = 0$. Suppose first that $P(Z) > 0$. Then, by the lemma, it is also true that $P(x) > 0$ in an interval $Z - \alpha < x < Z + \alpha$. But if the natural number k is such that $1/10^k < \alpha$, then by definition $Z - \alpha < b_k \leq Z \leq c_k < Z + \alpha$, and so we have $P(b_k) > 0$, which *contradicts* the choice of b_k. Suppose secondly that $P(Z) < 0$. By applying the lemma to the polynomial $-P(x)$, it can be shown this time that there must be a natural number k such that $P(c_k) < 0$, which is again a *contradiction* of the definition of c_k. We thus indeed have $P(Z) = 0$.

The lemma follows from a statement which can be written as follows:

$$P(x_0 + h) = P(x_0) + hP_1(x_0) + h^2 P_2(x_0) + \ldots + h^n P_n(x_0),$$

where $P_1, P_2, \ldots, P_n$ are polynomials which it is not necessary to give explicitly here.

If $|h| < 1$, we thus have

$$|P_1(x_0) + hP_2(x_0) + \ldots + h^{n-1}P_n(x_0)| < M,$$

where

$$M = 1 + |P_1(x_0)| + |P_2(x_0)| + \ldots + |P_n(x_0)|,$$

a number ≥ 1 and *independent* of h. Thus we have

$$P(x_0 + h) \geq P(x_0) - |h| \cdot |P_1(x_0) + hP_2(x_0) + \ldots + h^{n-1}P_n(x_0)|$$
$$\geq P(x_0) - M|h|.$$

If $P(x_0) > 0$, it follows that $P(x_0 + h) > 0$ for all numbers h such that $-1 < h < 1$ and

$$-\frac{P(x_0)}{2M} < h < \frac{P(x_0)}{2M}\,.$$

An analogous argument proves that, if

$$P(t) = a_0 t^n + a_1 t^{n-1} + \ldots + a_n\,,$$

where $a_0 > 0$, then there is a sufficiently large integer $A > 0$ such that $P(A) > 0$. In fact, for $t > 0$, we can write

$$P(t) = t^n \left(a_0 + \frac{1}{t}\left(a_1 + \frac{a_2}{t} + \ldots + \frac{a_n}{t^{n-1}} \right) \right).$$

If $t \geq 1$, we have

$$a_0 + \frac{1}{t}\left(a_1 + \frac{a_2}{t} + \ldots + \frac{a_n}{t^{n-1}} \right) \geq a_0 - \frac{1}{t}\left| a_1 + \frac{a_2}{t} + \ldots + \frac{a_n}{t^{n-1}} \right|$$

$$\geq a_0 - \frac{N}{t}\,,$$

where $N = 1 + |\,a_1\,| + |\,a_2\,| + \ldots + |\,a_n\,|$. If A is an integer such that $A > N/a_0$, we thus have

$$a_0 + \frac{1}{A}\left(a_1 + \frac{a_2}{A} + \ldots + \frac{a_n}{A^{n-1}} \right) \geq a_0 - \frac{N}{A} > 0\,,$$

and $P(a)$, being the product of A^n and a number > 0, is also a number > 0.

Stevin's method applies in an obvious way to an arbitrary function $f(t)$ for which $f(0) < 0$ and $f(A) > 0$ for a sufficiently large natural number $A > 0$ *provided that* the lemma proved above for a polynomial also holds for this function. This is the case for all the functions which we call *continuous*: these are the functions such that, for each point x of the interval on which they are defined, and for *each* number $\alpha > 0$, there exists a number $\beta > 0$ (depending on x and on α) such that

$$|\,f(x+h) - f(x)\,| < \alpha$$

whenever $-\beta < h < \beta$. In fact, if $f(x_0) > 0$, it suffices in the above definition to take the case where $\alpha = f(x_0)/2$ to ensure that

$$f(x_0 + h) - f(x_0) \geq -\frac{1}{2}f(x_0),$$

whence $f(x_0 + h) \geq f(x_0)/2 > 0$ whenever $-\beta < h < \beta$. It is shown that an equivalent formulation is that $\lim_{n \to \infty}(f(x + h_n) - f(x)) = 0$ for each sequence (h_n) tending to 0. Until the end of the eighteenth century, mathematicians implicitly assumed that all the functions which they studied were continuous.

4. Arguments by "Exhaustion"

"Exhaustion", described by Euclid (Book VII, 2), is based on the evaluation of the area of the triangle ACB formed by a chord and the tangents to a circle at the endpoints of the chord (Figure 29). This area is twice that of the area of triangle ACH, and this latter is

$$\frac{1}{2}AH \cdot CH = \frac{1}{2}AH^2 \tan\alpha = \frac{1}{2}\sin^2\alpha \tan\alpha = \frac{1}{2}\frac{\sin^3\alpha}{\cos\alpha}.$$

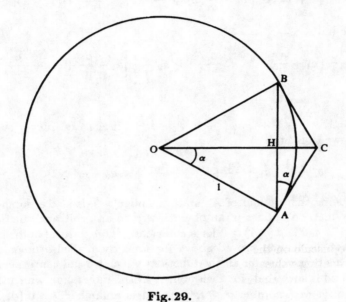

Fig. 29.

Euclid's proof that this area is reduced by more than half when the arc AB is divided into two thus amounts to the trigonometric inequality

$$(1) \qquad \frac{\sin^3\alpha}{\cos\alpha} < \frac{1}{4}\frac{\sin^3 2\alpha}{\cos 2\alpha} = 2\frac{\sin^3\alpha\cos^3\alpha}{\cos 2\alpha}$$

for $\alpha < \pi/4$. This is equivalent to $\cos 2\alpha < 2\cos^4\alpha$, and finally to

$$2\cos^4\alpha - 2\cos^2\alpha + 1 > 0,$$

and this follows from the obvious inequality

$$2\left(x^2 - \frac{1}{2}\right)^2 + \frac{1}{2} > 0.$$

Archimedes used Euclid's proposition to prove that the area of a circle is "equal to that of a right-angled triangle for which one side adjacent to the right-angle has the same length as the circumference and the other side is equal to the radius of the circle". Thus he did not hesitate to speak of the length of an arc of a curve (of which there was no question in Euclid's school), and was conscious that this necessitates new postulates: the essential one consists in saying that the length of an arc of a *convex* curve Γ contained in a triangle ABC, where BC is the chord with endpoints B and C on Γ (Figure 30), is equal to the length of a segment lying between BC and $AB + AC$.

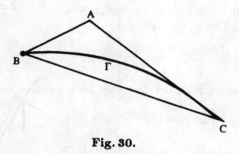

Fig. 30.

With the notations of §9, if l_n and L_n are the perimeters of the regular polygons P_n and Q_n of 2^n sides, respectively inscribed in and circumscribed around a circle of unit radius, then Archimedes could write the inequality as

$$l_n \leq 2\pi \leq L_n.$$

But in his opuscule on the length of a circle, he restricted himself to finding the approximating values given by the perimeters of regular polygons of 96 sides inscribed in and circumscribing the circle.

One would expect him to proceed as Euclid would, by proving the inequality

$$L_n - l_n \leq \frac{1}{2}(L_{n-1} - l_{n-1}),$$

which would give him 2π as the "limit" of the perimeters l_n and L_n.

We have here, in Figure 29,

$$(AB + BC) - AB = 2(\tan\alpha - \sin\alpha),$$

and the inequality (2) is a consequence of the trigonometric inequality

$$\tan\alpha - \sin\alpha \leq \frac{1}{4}(\tan 2\alpha - \sin 2\alpha)$$

for $\alpha < \pi/4$. But this inequality can be written

$$\frac{\sin\alpha\,(1 - \cos\alpha)}{\cos\alpha} \leq \frac{1}{4}\frac{\sin 2\alpha\,(1 - \cos 2\alpha)}{\cos 2\alpha} = \frac{\sin\alpha\,\cos\alpha\,(1 - \cos^2\alpha)}{\cos 2\alpha},$$

and after several elementary transformations, this reduces to

$$\cos 2\alpha \le \cos^2 \alpha \left(1 + \cos \alpha\right),$$

which is to say, finally, that

$$\cos^3 \alpha - \cos^2 \alpha + 1 \ge 0.$$

Certainly we have $x^3 - x^2 + 1 \ge 0$ whenever $0 \le x \le 1$.

However trigonometric calculations did not become current until after Ptolemy (second century A.D.), and a purely geometric proof of the inequality (3) must have seemed too difficult to Archimedes.

As to the calculation of the primitive

$$\int_0^x t^m \, dt$$

for a natural number m, it comes down to the evaluation of the sum

(4) $$\frac{x^{m+1}}{n^{m+1}} \left(1^m + 2^m + \ldots + (n-1)^m\right),$$

as Archimedes had seen in the case where $m = 2$.

The parenthesis in (4) is equal to the value $S_m(n)$, where $S_m(t)$ is a polynomial of degree $m + 1$ with rational coefficients. This is proved by writing that this polynomial

$$S_m(t) = a_0 t^{m+1} + a_1 t^m + \ldots + a_m t$$

is such that

$$S_m(n+1) - S_m(n) = n^m.$$

From this, we may determine the coefficients a_j (cf. Chapter IV, Appendix 4). In particular, we obtain $a_0 = 1/(m+1)$. The sum (4) is thus a sum of $x^{(m+1)}/(m+1)$ and of m terms each of which tends to 0 as n increases indefinitely. Fermat's result follows from this.

5. Applications of Elementary Algorithms of the Integral Calculus

For benefit of those readers who have a familiarity with the rudiments of the infinitesimal calculus, I would like to give several examples showing what can be pulled out of some elementary algorithms of the integral calculus which Hardy judged "dull and boring" (Chapter II, §1). To simplify, we shall suppose, as did the mathematicians of the eighteenth century, that the functions considered are continuous and piecewise monotone (i.e., their interval of definition can be divided into a finite number of subintervals, on each of which the function is either increasing or decreasing or constant). These algorithms are three in number:

1) If $f(x) \ge 0$ on $[a, b]$, then we have

$$\int_a^b f(x)\,dx \geq 0,$$

and we can only have

$$\int_a^b f(x)\,dx = 0$$

if $f(x) \equiv 0$ on $[a, b]$. It can be deduced that, for two arbitrary functions f and g such that $|f(x)| \leq M$ on $[a, b]$, we have

(1)
$$\left| \int_a^b f(x)g(x)\,dx \right| \leq M \int_a^b |g(x)|\,dx.$$

2) For two differentiable functions u and v on $[a, b]$, we have

(2)
$$\int_a^b u(x)v'(x)\,dx = u(b)v(b) - u(a)v(a) - \int_a^b u'(x)v(x)\,dx$$

("integration by parts").

3) If ϕ is defined and differentiable on $[a, b]$ and f is defined on an interval containing all the values taken by ϕ, then we have

$$\int_a^b f(\phi(t))\phi'(t)\,dt = \int_{\phi(a)}^{\phi(b)} f(x)\,dx$$

("change of variables").

The formulae (2) and (3) are only other ways of writing the formulae which give the derivatives of a product and of a composition of functions.

Mathematicians of the eighteenth century used these formulae above all to obtain the primitives of the "elementary" functions which themselves are "elementary" functions, but they have many other uses. (For these mathematicians, the notion of "elementary" function was not very precise. Ordinarily these functions were those obtained by repeated compositions of algebraic operations and of "usual" functions, such as e^x, $\log x$, $\sin x$, etc. For example,

$$(e^{x^2} + \sin x)^{1/3}$$

must be considered as an "elementary" function. In the nineteenth century, Liouville gave more precise definitions and was able to prove that the primitives of certain "elementary functions", such as e^x/x, are not "elementary".)

One of the most important of the formulae of analysis, *Taylor's formula* for a function which is $n+1$ times differentiable, results from repeated applications of formula (2):

$$\int_a^x \frac{(x-t)^n}{n!} f^{(n+1)}(t)\,dt = -\frac{(x-a)^n}{n!} f^{(n)}(a) + \int_a^x \frac{(x-t)^{n-1}}{(n-1)!} f^{(n)}(t)\,dt$$

$$\cdots\cdots\cdots\cdots\cdots\cdots\cdots\cdots\cdots\cdots\cdots\cdots$$

$$\int_a^x (x-t)f''(t)\,dt = -(x-a)f'(a) + \int_a^x f'(t)\,dt$$

$$\int_a^x f'(t)\,dt = f(x) - f(a),$$

which gives, by adding the various terms,

$$f(x) = f(a) + \frac{(x-a)}{1!} f'(a) + \ldots + \frac{(x-a)^n}{n!} f^{(n)}(a) + \int_a^x \frac{(x-t)^n}{n!} f^{(n+1)}(t)\,dt.$$

Here are two other examples:

I) In probability theory, the error that is made when the area situated under the "bell-shaped curve" for large values of the variable is neglected must be evaluated. That is to say the integral

$$\int_x^\infty e^{-t^2}\,dt$$

must be calculated. This function of x is not "elementary", but, on integrating by parts, it can be written

$$\int_x^\infty e^{-t^2}\,dt = \int_x^\infty \frac{1}{2t} \cdot 2t\,e^{-t^2}\,dt = \frac{e^{-x^2}}{2x} - \int_x^\infty \frac{e^{-t^2}}{2t^2}\,dt,$$

and since, for $x \le t$ we have

$$1 \le 1 + \frac{1}{2t^2} \le 1 + \frac{1}{2x^2},$$

we obtain

$$\frac{e^{-x^2}}{2x\left(1 + \dfrac{1}{2x^2}\right)} \le \int_x^\infty e^{-t^2}\,dt \le \frac{e^{-x^2}}{2x}.$$

II) In numerous questions of analysis, of mathematical physics, and of number theory, we have functions of the form

$$e^{if(t)} = \cos f(t) + i \sin f(t),$$

where f is a real-valued function, often qualified as the "phase function". The point $e^{if(t)}$ is on the circle of radius 1 and "turns" more or less quickly on the circle according as $f(t)$ varies more or less rapidly. Since $|e^{if(t)}| = 1$, the obvious upper bound given by (1) for the integral

$$\left| \int_a^b e^{if(t)}\,dt \right|$$

is only the estimate $b - a$. In fact, this is only attained when f is constant. If $f(t) = ct$, we have

$$\int_a^b e^{ict}\, dt = \frac{1}{ic}(e^{icb} - e^{ica}),$$

whose absolute value may be very large when c is very small. In this latter case, the second derivative $f''(t)$ is identically zero. It is remarkable that the absolute value of the integral may be bounded above when one only knows a lower bound of $|f''(t)|$. If $f''(t) \ge \lambda > 0$ on $[a, b]$, then we have

(4)
$$\left| \int_a^b e^{if(t)}\, dt \right| \le \frac{8}{\sqrt{\lambda}}$$

(van der Corput's inequality).

We examine several cases.

1) $b - a \le 2/\sqrt{\lambda}$; then, by (1), we have

$$\left| \int_a^b e^{if(t)}\, dt \right| \le \frac{2}{\sqrt{\lambda}}.$$

2) $b - a > 2/\sqrt{\lambda}$ and $f'(t) \ge 0$ on $[a, b]$. Then we have (since f' is increasing) that

$$f'(t) \ge \lambda \cdot \frac{2}{\sqrt{\lambda}} = 2\sqrt{\lambda} \quad \text{for} \quad a + \frac{2}{\sqrt{\lambda}} \le t \le b.$$

We can then write by (2)

$$\int_{a+2/\sqrt{\lambda}}^b e^{if(t)}\, dt = \int_{a+2/\sqrt{\lambda}}^b \frac{1}{if'(t)} \cdot if'(t)\, e^{if(t)}\, dt$$

$$= \frac{e^{if(b)}}{if'(b)} - \frac{e^{if(a+2/\sqrt{\lambda})}}{if'\left(a + \frac{2}{\sqrt{\lambda}}\right)} + \frac{1}{i} \int_{a+2/\sqrt{\lambda}}^b \frac{f''(t)}{(f'(t))^2}\, e^{if(t)}\, dt,$$

and, by (1), it follows that

$$\left| \int_{a+2/\sqrt{\lambda}}^b \frac{f''(t)}{(f'(t))^2}\, e^{if(t)}\, dt \right| \le \int_{a+2/\sqrt{\lambda}}^b \frac{f''(t)}{(f'(t))^2}\, dt$$

$$= \frac{1}{f'\left(a + \frac{2}{\sqrt{\lambda}}\right)} - \frac{1}{f'(b)}$$

$$\le \frac{1}{\sqrt{\lambda}},$$

from which it is concluded that

$$\left| \int_a^b e^{if(t)}\, dt \right| \le \left| \int_a^{a+2/\sqrt{\lambda}} e^{if(t)}\, dt \right| + \left| \int_{a+2/\sqrt{\lambda}}^b e^{if(t)}\, dt \right| \le \frac{4}{\sqrt{\lambda}}.$$

3) *General case.* Since f'' does not change sign, f' can only be zero at one point c such that $a \leq c \leq b$. Thus, by the preceding case,

$$\left| \int_c^b e^{if(t)} \, dt \right| \leq \frac{4}{\sqrt{\lambda}} \, .$$

On the other hand, by (3), we have

$$\int_a^c e^{if(t)} \, dt = \int_{-c}^{-a} e^{ig(t)} \, dt \, ,$$

where $g(t) = f(-t)$. Since $g''(t) = f''(-t)$, we have $g''(t) > \lambda$ on the interval $[-c, -a]$, and $g'(t) = -f'(-t) \geq 0$ on this interval. By the preceding case, we thus have

$$\left| \int_a^c e^{if(t)} \, dt \right| = \left| \int_{-c}^{-a} e^{ig(t)} \, dt \right| \leq \frac{4}{\sqrt{\lambda}} \, ,$$

which completes the proof of (4).

IV Some Problems of Classical Mathematics

In Chapter V we shall attempt to show how, starting from classical objects and methods, mathematicians of the nineteenth century were *obliged* to invent *new mathematical objects* and *new methods* in order to make further progress.

In order that this necessity should be understood, I think it may be of use, in this chapter and the next, briefly to go over the history of some of the problems attacked by mathematicians, with widely varying degrees of success. In this chapter, however, we shall deal only with those problems which were eventually solved by using none but the "classical" objects which we have already discussed: numbers and geometric figures. In order to remain intelligible to the readers whom I am addressing, I shall have to leave aside the problems connected with what is known as *mathematical analysis*, that is, the development and the multiple uses of the concepts of the calculus (see Chapter III, §9); however, these problems are the most numerous, and have been the subject of the most spectacular advances. Even for certain problems which can be stated in perfectly elementary terms, especially imaginative mathematicians have managed to obtain partial or complete solutions by bringing in concepts or techniques drawn from analysis which seem to have nothing to do with the question in hand. I can hardly do more than allude to these methods, which amaze mathematicians, making them feel the profound and often mysterious unity of mathematics, and in speaking of which they do not hesitate to use the term "beauty".

1. Intractable Problems and Sterile Problems

A. Perfect Numbers

Certain problems, still unsolved today, go back to antiquity. The philosophico-mystical concepts which the Pythagoreans associated with natural numbers led them to observe that certain natural numbers have the property of being equal to the *sum of their divisors*[1]. For example,

$$28 = 14 + 7 + 4 + 2 + 1.$$

---

[1] In this statement a number n is not counted among its own divisors.

They called these the *perfect numbers*, and the problem was to find them *all*. A testimony of the remarkable advances achieved by Greek arithmetic is that Euclid was able to prove (*Elements*, Book IX, 36) that a number of the form

$$N = 2^{p-1}(2^p - 1) \qquad (\text{for} \quad p \geq 2)$$

is perfect *provided* that the number $2^p - 1$ is *prime*. In fact, the only divisors of N are

$$1, \ 2, \ 2^2, \ ..., \ 2^{p-1}$$

and

$$2^p - 1, \ 2(2^p - 1), \ ..., \ 2^{p-2}(2^p - 1)$$

since $2^p - 1$ is prime. But then the sum of the first set of divisors is $2^p - 1$ and that of the second set is $(2^{p-1} - 1)(2^p - 1)$, and therefore the sum of all the divisors is

$$2^p - 1 + (2^{p-1} - 1)(2^p - 1) = 2^{p-1}(2^p - 1) = N \, .$$

In the eighteenth century Euler proved that these numbers are the only perfect *even* numbers. But this result immediately gives rise to two problems:

1) Are there any perfect *odd* numbers?

2) For which natural numbers $p \geq 2$ is the number $2^p - 1$ prime?

The first question is still unanswered today, although various calculations tend to suggest that the answer is negative. As for the second, it is at once clear that the natural number p itself must be prime, since if $p = mn$ with $m > 1$ and $n > 1$, we have

$$2^{mn} - 1 = (2^n - 1)(1 + 2^n + 2^{2n} + ... + 2^{(m-1)n}) \, .$$

Numbers of the form $2^p - 1$ where p is prime, are called *Mersenne numbers*. With the help of very laborious calculations "by hand", it has been verified that for $p \leq 127$, the only Mersenne numbers which are prime are those corresponding to the values

$$2, \quad 3, \quad 5, \quad 7, \quad 13, \quad 17, \quad 19, \quad 31, \quad 61, \quad 89, \quad 107, \quad 127$$

of p.

With the advent of computers, it has been possible to extend this list, and nearly every year a new Mersenne prime number is discovered. In 1985 the largest was

$$2^{216\,091} - 1,$$

which has $65\,050$ digits, and was the largest known prime number. But it is still not known if there is an *infinity* of Mersenne numbers which are prime.

B. Fermat Numbers

Fermat seems to have been the first to put to himself an analogous problem for the numbers $2^k + 1$. Since the polynomial $T^k + 1$ is divisible by $T + 1$

when k is *odd*, the only numbers $2^k + 1$ which can be prime are the numbers (called "Fermat numbers")

$$2^{2^n} + 1.$$

Fermat had verified that

$$2^{2^0} + 1 = 3, \quad 2^{2^1} + 1 = 5, \quad 2^{2^2} + 1 = 17,$$
$$2^{2^3} + 1 = 257, \quad 2^{2^4} + 1 = 65\,537$$

were primes, and he thought that *all* "Fermat numbers" $2^{2^n} + 1$ were too. But Euler showed that

$$2^{2^5} + 1 = 641 \times 6\,700\,417,$$

and up to now *no* other Fermat number which is a prime has been found.

What characterizes the preceding problems is not that there are points which have still not been cleared up in the study of them – that is so in *all* problems – it is that it has never been possible to advance one step in them. Of course it would be easy to enumerate indefinitely statements of analogous problems which are just as intractable. But why have those which we have cited been kept alive in mathematicians' memories? "Perfect" numbers take us back to the dawn of arithmetic and to the wonder of our ancestors; while Mersenne and Fermat numbers are connected with famous names, and recently they have come in a strange way into the theory of finite simple groups (Chapter V, §4, B). Further, Gauss's first great mathematical discovery, at the age of nineteen, was that a regular polygon with prime number p sides can be constructed with ruler and compass *if and only if p is a Fermat number*.

Moreover, one must never despair: problems which fifty years ago seemed out of reach have since been solved (see Appendix 4).

C. The Four-Colour Problem

This is a problem first posed about 1850 which immediately fascinated non-mathematicians because it can be stated so simply, and yet is so difficult to solve. A geographical map shows N countries, each of which has several neighbours (Figure 31). The problem is to find out if four *colours* are enough to colour each of the N countries so that no two adjacent countries have the same colour.

This is a problem in what is called *combinatorics*, in which, for each value of N, all possible positions of N countries relative to the others must be examined. Over more than a century much ingenuity has gone into dealing with larger and larger numbers N (a number such as 50 already requires a good deal of work). Notably, typical positions have been studied which could be used to reduce the problem for N countries to a problem for a smaller number N'. Finally, in 1977, Appel and Haken succeeded in showing that it was sufficient to examine a *finite* number M of well-determined configurations to be certain that the use of four colours was possible for *every* configuration. But this number M was over $1\,800$, and there was no question of its being

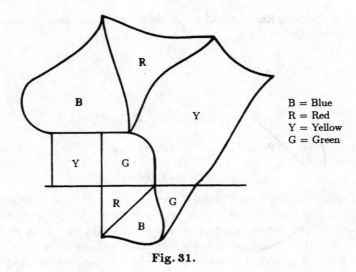

Fig. 31.

possible to examine all these cases "by hand": it was necessary to call in the help of a powerful computer, which, after a year's work, managed to prove the theorem. It seems that several independent verifications have been made, and we can therefore consider that the four-colour problem has been solved, while hoping that one day another method will produce a less bulky proof.

All the same, what we can say of this gargantuan labour is that it seems disproportionate to the final result. We shall see later (§2, B) that, in the theory of prime numbers, there remains an unproved conjecture, the Riemann Hypothesis. Its interest for mathematicians arises from the fact that dozens of other results depend on it: they would be proved *if* the Riemann Hypothesis were proved. There is, to my knowledge, nothing of this kind for the four-colour problem. It is an isolated result, which has not been used to prove other theorems in "combinatorics", and the same is true of the method used to prove it. Of course we cannot speak for the future, but we can say that, until a new result appears, this is a problem *without issue*.

D. Problems of Elementary Geometry

The same conclusion applies, for the same reasons, to most of the problems in elementary plane geometry which have entranced generations of teachers and enthusiasts of mathematics. For example:

1) *Castillon's Problem.* Given a circle Γ and any three points P, Q, and R not on the circle, the task is to find a triangle ABC inscribed in the circle Γ and with sides (which may be extended) passing through P, Q, and R, respectively (Figure 32).

2) *Malfatti's Problem.* This is to find three circles, each of which touches the two others, and which also touches two sides of a given triangle ABC (Figure 33).

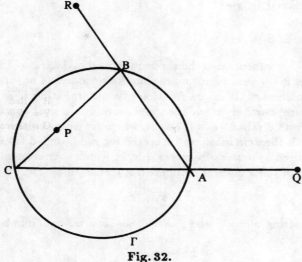

Fig. 32.

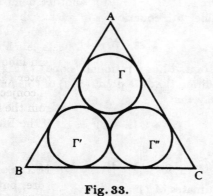

Fig. 33.

3) *Compass Geometry.* Here the task is to find constructions made with compass *alone* which allow one to carry out all the constructions which can be made with a ruler and a compass.

The solutions to these problems and to many others, as they are set out in works on geometry, have perhaps helped to give the impression of a complete science in which there was nothing more to be discovered. Paradoxically, it is the geometrical problems which are generally not mentioned in these works, problems which it has not been possible to solve "with ruler and compass", which have been linked to substantial progress in the understanding of mathematics (Chapter V, §1).

2. Prolific Problems

A. The Sums of Squares

In Chapter II, §5, we mentioned the theorem proved by Lagrange in 1770: each natural number is the sum of not more than four squares of natural numbers. In fact, the problems of the representation of a natural number as the sum of squares of natural numbers made its appearance in classical times, no doubt in connection with Pythagoras's theorem, which we now explain.

Pythagoras's theorem leads to the search for right-angled triangles whose sides are measured by *natural numbers a, b, c*, which is equivalent to seeking all the systems of three natural numbers a,b, c such that

$$a^2 = b^2 + c^2 \,.$$

Equivalently, setting $x = b/a$ and $y = c/a$, we seek *rational* numbers x and y such that

$$x^2 + y^2 = 1 \,.$$

This is the equation of a circle C, centre O and radius 1, and the point $A = (1,0)$ is on this circle. If $M = (x,y)$ is another point of C with rational coordinates, then the line with equation

$$Y = t(X - 1)$$

which joins M to the point A has a rational "slope" t (Figure 34).

If we write the condition that the point (x,y) of this line satisfy the equation of the circle, namely that

$$x^2 + t^2(x-1)^2 = 1 \,,$$

then we obtain

$$(x - 1)(x + 1 + t^2(x - 1)) = 0 \,,$$

and from this the coordinates of M are

$$x = \frac{t^2 - 1}{1 + t^2}, \qquad y = \frac{-2t}{1 + t^2} \,.$$

If we take for t any rational number, then we obtain *all* the rational solutions of $x^2 + y^2 = 1$.

Diophantus already knew that a number of the form $4k + 3$ cannot be the sum of two squares, that a number of the form $8k + 7$ cannot be the sum of three squares[2], and it is possible that he conjectured Lagrange's theorem. In any case, this conjecture was voiced in the seventeenth century by Bachet de

---

[2] Every number has one of the forms $4k$, $4k + 1$, $4k + 2$, or $4k + 3$. Taking the square of the number, it is evident that all the numbers obtained are of the form $4m$ or $4m + 1$, and that the sum of two of these numbers can never be of the form $4m + 3$. For three squares the reasoning is the same, except that the squares of numbers of the form $8k + h$ are considered, with $0 \leq h \leq 7$.

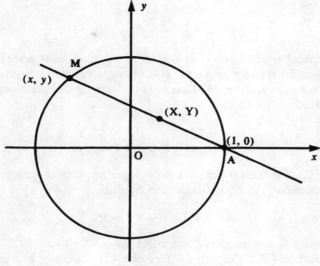

Fig. 34.

Méziriac, and Fermat claimed to have proved it. He also claimed that each *prime* number of the form $4k+1$ is *the sum of two squares in only one way*, but his proofs have not been found among his papers. The second result was proved by Euler, who also succeeded in showing that each natural number is the sum of four squares of *rational* numbers. On the other hand Legendre published that each natural number which *is not* of the form $4^m(8k+7)$, is the sum of at most three squares of natural numbers. His proof was not altogether convincing, and it was completed by Gauss.

What is remarkable is that these theorems, far from making mathematicians turn aside from these questions which had apparently been solved, only gave rise to a host of new problems, the study of which constitutes one of the most extensive and most flourishing sections of number theory; one of those, too, in which the connections with other areas of mathematics, such as analysis or geometry, have been the most surprising and the most fertile. Unfortunately, the proofs either call for advanced techniques in analysis or else (like those of Fermat and Euler) are elementary, but too long and complicated to be reproduced here (see Appendix 3). We must therefore limit ourselves to a few statements of problems and their solutions.

I) Apart from the sums of two or four squares, Fermat also studied the natural numbers which can be written in the form $a^2 + Nb^2$, where a and b are natural numbers, in the case where $N = 2$ or $N = 3$. Euler proved and generalized the theorems which Fermat had stated on this subject without giving the proofs.

It was Lagrange who, starting with these results, inaugurated a new approach destined to lead to amazing advances. He considered in a general way equations of the form

(1) $$ax^2 + bxy + cy^2 = n,$$

where a, b, c, and n are integers (positive or negative) and where solutions (x, y) are sought for which x and y are (positive or negative) integers. (We say simply that these are *integer* solutions of (1)). Lagrange's idea was to make a "change of variables" in (1)

(2) $$\begin{cases} x = \alpha X + \beta Y \\ y = \gamma X + \delta Y, \end{cases}$$

where α, β, γ, and δ are *integers*. It is obvious that this change of variables transforms equation (1) into an analogous equation

(3) $$AX^2 + BXY + CY^2 = n,$$

where A, B, and C are integers (depending on α, β, γ, and δ)[3]. If we have an integral solution (X, Y) of (3), the formulae (2) obviously give an integral solution (x, y) of (1). In general, however, *all* the integral solutions of (1) are not to be obtained in this way, because, if we resolve the system of two linear equations in X and Y, we derive from (2) that

(4) $$\begin{cases} X = \dfrac{\delta}{\alpha\delta - \beta\gamma}\, x - \dfrac{\beta}{\alpha\delta - \beta\gamma}\, y \\ Y = \dfrac{\alpha}{\alpha\delta - \beta\gamma}\, y - \dfrac{\gamma}{\alpha\delta - \beta\gamma}\, x, \end{cases}$$

and, if x and y are integers, then X and Y are generally only *rational* numbers. This happens unless we have

(5) $$\alpha\delta - \beta\gamma = \pm 1.$$

Lagrange therefore confines himself to the changes of variable (2) which satisfy condition (5), and he then says that the polynomials $ax^2 + bxy + cy^2$ and $AX^2 + BXY + CY^2$ are *equivalent quadratic forms*. Later Gauss was to show that it is preferable to restrict this concept, using the term *properly equivalent* for two quadratic forms which are obtained from one another by a change of variables (2) such that $\alpha\delta - \beta\gamma = 1$. It can then also be said that the quadratic

---

[3] It is such changes of variables (with *arbitrary* coefficients) which makes it possible to rewrite equation (1), which represents a conic, in the form

$$AX^2 + BY^2 = 1.$$

But if it is required that α, β, γ, and δ be *integers* and satisfy (5), it is not always possible to choose them so as to ensure that, in (3), the "rectangular term" BXY is zero. For example, for the form $x^2 + xy + y^2$, we have $B = 2(\alpha\beta + \gamma\delta) + \alpha\delta + \beta\gamma$, and since $\alpha\delta - \beta\gamma = \pm 1$, the relation $B = 0$ becomes $2(\alpha\beta + \beta\gamma + \gamma\delta) = \pm 1$, which is absurd when α, β, γ, and δ are integers.

forms such that any two are properly equivalent constitute one and the same *class*.

II) For a quadratic form $ax^2 + bxy + cy^2$, the number $D = b^2 - 4ac$ is called the *discriminant* of the form: it is an integer. All the forms of the same class have the same discriminant, as is shown by an immediate calculation based on (2); but conversely the forms with a given discriminant D do not always belong to the same class. Nevertheless, the number $h(D)$ of classes with discriminant D is *finite*: it was explicitly determined as a function of D by Dirichlet. In the twentieth century it has been shown that $h(D)$ increases indefinitely when D tends to $-\infty$, as was conjectured by Gauss. On the other hand, he also conjectured that when D tends to $+\infty$, $h(D)$ takes the value 1 infinitely many times, but it is not known how to prove or disprove this conjecture.

It is not known, in general, what criteria are necessary and sufficient to ensure that an integer n is such that the equation (1) has at least one integer solution (x, y). Lagrange and Gauss could only give one criterion which brings in only the discriminant D and the integer n to ensure that (1) has at least one integer solution (x, y) for *at least one form with discriminant D*. But if there are several classes corresponding to the discriminant D, it is not possible to say for which class there will be an integer solution of (1). On the other hand, Lagrange showed how, given that there exists *one* solution of (1), all the others can be found: there is a finite number of them if $D < 0$, but infinitely many if $D > 0$.

Up to now all these results have referred only to classical objects and methods, but we shall see in Chapter V, §3, CII), how they have been one of the main sources of modern number theory, and have thus contributed to the introduction of new mathematical objects: *fields*, *rings* and *modules*.

III) Lagrange's theorem on the sums of four squares at once gives rise to a new problem: *how many* integral solutions (x_1, x_2, x_3, x_4) are there of the equation

(6) $$x_1^2 + x_2^2 + x_3^2 + x_4^2 = n\,?$$

The answer was given by Jacobi, and it is connected with the arithmetical properties of n in an unexpected way. For each exponent $k \geq 1$ we designate by $\sigma_k(n)$ the *sum of the* kth *powers of the divisors of* n[4]. Then the number of solutions of (6) is $8\sigma_1(n)$ if n is odd. If on the other hand $n = 2^r m$, where m is odd and $r \geq 1$, then the number of solutions is $24\sigma_1(m)$.

Naturally they were not going to stop there now they were so well launched, and the problem was generalized: for *each* natural number $m > 4$, find the number $R_m(n)$ of integer solutions $(x_1, x_2, ..., x_m)$ of

(7) $$x_1^2 + x_2^2 + ... + x_m^2 = n\,.$$

---

[4] Here a number is considered as one of its own divisors. Thus the definition given above of a perfect number is equivalent to $\sigma_1(n) = 2n$.

Here we have no result which is complete for all cases. If $m = 8$, we have $R_8(n) = 240\sigma_3(n)$, and, if $m = 16$, then $R_{16}(n) = 480\sigma_7(n)$. In general, if m is a multiple of 8, we have an "approximate expression" for $R_m(n)$:

$$(8) \qquad \frac{m}{B_k}\,\sigma_{2k-1}(n),$$

where B_k is the kth Bernoulli number (Appendix 4), with $k = m/4$. "Approximate expression" means that

$$\varphi_m(n) = R_m(n) - \frac{m}{B_k}\,\sigma_{2k-1}(n)$$

is very small in relation to $\sigma_{2k-1}(n)$; to be precise, we obviously have

$$\sigma_{2k-1}(n) > n^{2k-1},$$

and it is proved that

$$|\varphi_m(n)| \le C_m n^k,$$

where C_m is a constant independent of n.

These results still require only the classical analysis of the nineteenth century for methods of proof, but they were to be among the sources of the new concept of *group* (see Chapter V, 2, C).

The next stage is to consider instead of (7) an equation

$$(9) \quad a_1 x_1^2 + a_2 x_2^2 + \dots a_m x_m^2 +$$
$$+\, b_{12} x_1 x_2 + b_{13} x_1 x_3 + \dots + b_{m-1,m} x_{m-1} x_m = n\,,$$

a generalization of (1) in which the coefficients a_k and b_{ij} are integers. The theory follows a course analogous to that of equation (1), but it is much more complicated and requires more powerful analytic methods.

IV) A further step in the generalization of equation (6) was taken by Waring even before Lagrange's proof: he asked if there exists for *each* $k \ge 3$, an integer $G(k)$, depending only on k, such that, for *every* integer n, there is an integral solution $(x_1, x_2, \dots, x_{G(k)})$ of the equation

$$(10) \qquad\qquad x_1^k + x_2^k + \dots + x_{G(k)}^k = n\,.$$

Here it was not until 1910 that the existence of $G(k)$ was proved by Hilbert for each $k \ge 3$. Since then, the question has been the subject of some very deep research, in the course of which some approximate estimates have been made of the number of solutions of (10). As for the arithmetical theory of the "forms of degree k", which would generalize the theory of quadratic forms, it is still, despite many endeavours, at the embryonic stage. It is the same with systems of several equations of arbitrary degree.

B. The Properties of Prime Numbers

To my knowledge, there was no ancient civilization other than the Greek one in which anyone thought, before the fifth century B.C., of breaking down a

natural number into its prime factors. This factorization, which we now write
as

(11) $$n = p_1^{k_1} p_2^{k_2} ... p_r^{k_r} ,$$

where $p_1, p_2, ..., p_r$ are prime numbers and the k_i are exponents at least equal
to 1, does not occur explicitly in Euclid's work for lack of an adequate notation.
He nevertheless proves the following three properties (expressed in modern
language):

a) Each natural number is either a prime or divisible by a prime number
(Book VII, 31).

b) If p is a prime, the only divisors of p^m are the numbers p^r with $r < m$
(Book IX, 13).

c) If a prime number divides a product ab of two natural numbers and
does not divide a, then it divides b (Book VII, 32).

From these properties it is easy, reasoning by induction, to establish the
existence and the uniqueness of the factorization (11).

Let us recall that we cited in Chapter II, §5, the most beautiful theorem
of Greek arithmetic, the fact that there exists an *infinity* of prime numbers.
The proof given by Euclid is very simple (see Hardy, [7], p. 28 and Appendix
I), but I prefer to give another, due to Euler, because it opened the way to
what is called the "analytic theory of prime numbers". It can in any case be
presented without using anything but elementary algebra.

It is an argument "by contradiction", in which it is supposed that

$$p_1 = 2, \quad p_2 = 3, \quad p_3 = 5, \quad ..., \quad p_r$$

(arranged in increasing order) are the *only* prime numbers; then from this is
deduced an absurd conclusion. By (11), *every* natural number n can be broken
down into the product on the right-hand side in one way only, admitting this
time that certain of the exponents k_i can be 0 (the factor $p_i^{k_i}$ being then
replaced by 1). We take an *arbitrarily large* natural number N, and consider
the product

(12) $$S_{N,r} = \left(1 + \frac{1}{2} + \frac{1}{2^2} + ... + \frac{1}{2^N}\right) \times$$

$$\times \left(1 + \frac{1}{3} + \frac{1}{3^2} + ... + \frac{1}{3^N}\right) ... \left(1 + \frac{1}{p_r} + \frac{1}{p_r^2} + ... + \frac{1}{p_r^N}\right) .$$

In order to calculate this product, it is necessary to take a term in each
parenthesis and compute the product of these, then to take the sum of all
these partial products. Now these partial products are written $1/n$, where n
is of the form shown in (11), but with the restriction that $0 \leq k_i \leq N$
for *all* the exponents. By the uniqueness of the factorization, all these partial
products are *different*. The essential point is that *all the natural numbers n
between* 1 *and* 2^N appear (once only, by what has just been said) in a partial

product $1/n$. In fact, if $1 \leq n \leq 2^N$, then *none* of the exponents $k_2, k_3, ..., k_r$ in the factorization (11) can be *greater than* $N - 1$, for otherwise the number n would be equal to at least 3^N, contradicting the supposition that $n \leq 2^N$. Thus $1/n$ does appear as one of the partial products in the expression for $S_{N,r}$. Of course, there are *other* partial products in $S_{N,r}$, but what we have shown is that we have

$$(13) \qquad S_{N,r} \geq 1 + \frac{1}{2} + \frac{1}{3} + \frac{1}{4} + ... + \frac{1}{2^N - 1} + \frac{1}{2^N}.$$

The sum on the right-hand side is not easy to evaluate, but it can be replaced by a smaller number in the following manner: the terms are grouped in partial sums stopping at the powers of $1/2$:

$$1 + \frac{1}{2} + \left(\frac{1}{3} + \frac{1}{4} \right) + \left(\frac{1}{5} + \frac{1}{6} + \frac{1}{7} + \frac{1}{8} \right) + ...$$
$$+ \left(\frac{1}{2^{N-1} + 1} + \frac{1}{2^{N-1} + 2} + ... + \frac{1}{2^N} \right).$$

The first parenthesis has two terms which are at least equal to $1/4$, thus it is $\geq 1/2$. The second has four terms at least equal to $1/8$, thus it is $\geq 4/8 = 1/2$. Continuing in this way, we see that *each* parenthesis is $\geq 1/2$, because it has 2^{k-1} terms, and all these are $\geq 1/2^k$. Since there are $N - 1$ parentheses, we have finally that

$$(14) \qquad S_{N,r} \geq 1 + \frac{1}{2}N.$$

It is possible, however, to express $S_{N,r}$ otherwise, by the formula giving the sum of a geometric progression,

$$1 + a + a^2 + ... + a^N = \frac{(1 - a^{N+1})}{(1 - a)},$$

whence

$$(15) \qquad S_{N,r} = \frac{\left(1 - \frac{1}{2^{N+1}} \right) \left(1 - \frac{1}{3^{N+1}} \right) \cdots \left(1 - \frac{1}{p_r^{N+1}} \right)}{\left(1 - \frac{1}{2} \right) \left(1 - \frac{1}{3} \right) \cdots \left(1 - \frac{1}{p_r} \right)}.$$

If all the factors of the numerator are replaced by 1, then we have the inequality

$$(16) \qquad S_{N,r} \leq \frac{1}{\left(1 - \frac{1}{2} \right) \left(1 - \frac{1}{3} \right) \cdots \left(1 - \frac{1}{p_r} \right)} = A_r,$$

and the right-hand side *no longer depends on* N. Comparing (14) and (16) we obtain

$$1 + \frac{1}{2}N \le A_r,$$

whence $N \le 2(A_r - 1)$, and since N is arbitrarily large, we have arrived at a contradiction.

Euler's idea, which was the germ of all later progress, was to replace the inverses $1/p_j$ of the prime numbers in (12) by a *power* $(1/p_j)^m$, where the exponent m is greater than 1 (but not necessarily a natural number). Unfortunately, the formula which is thus obtained cannot be described except by means of the concepts of analysis, series and infinite products, which we cannot use here (see Appendix 2).

Once the fact that there is no end to the sequence of prime numbers has been established, it is at least possible to compile *tables* giving the primes below a certain number. Quite early there were tables going up to $3 \cdot 10^6$ (three million), and computers can do much better. The earliest known method for constructing these tables is what is known as "Eratosthenes' sieve". To find the prime numbers $\le x$, we write the sequence of *all* natural numbers $2, 3, 4, 5, ..., x$; we exclude the multiples of 2 from 4 onwards, then the multiples of 3 from 6 onwards, the multiples of 5 from 10 onwards, and so on: to be precise, after the kth. operation the $k + 1$ smallest numbers not excluded are primes, and, if p_{k+1} is the largest of them, then the $(k + 1)$th. operation consists in excluding the multiples of p_{k+1} from $2p_{k+1}$ onwards. We may stop at the last prime number p_r which is $\le \sqrt{x}$. In fact, if a natural number m is such that $\sqrt{x} < m \le x$ and has not been excluded, then it cannot be a product ab with $a > 1$ and $b > 1$, since at least one of the numbers a and b would be $\le \sqrt{x}$, and hence divisible by one of the primes already found, and m would have to have been excluded. Thus, all the numbers not excluded and $> \sqrt{x}$ are primes.

There are more powerful methods for compiling tables of prime numbers, but Eratosthenes' sieve may have been the process which prompted Euler to consider the number A_r occurring in formula (16). In fact, for a prime number $p \le \sqrt{x}$, there are

$$\left(1 - \frac{1}{p}\right) x$$

numbers such that $1 \le m \le x$ and which *are not* multiples of p. If there were not in "Eratosthenes' sieve" numbers which are excluded several times, there would be approximately

$$(17) \qquad \left(1 - \frac{1}{2}\right)\left(1 - \frac{1}{3}\right) ... \left(1 - \frac{1}{p_r}\right) x$$

prime numbers less than x, where p_r is the largest number $\le \sqrt{x}$. The multiple of x in (17) is precisely the denominator of the number A_r. Naturally this number depends on x, and Euler was able to show that it tends towards 0 when x increases indefinitely. If, however, we examine a table of primes we find that their distribution is extremely irregular: a large number of primes

p are known such that $p + 2$ is also a prime[5] (they are said to form pairs of "twin" primes), and it is even suspected that there are infinitely many of these, although up to now there is no known way of proving this. On the other hand there are in the sequence of integers "holes" as large as you like in which there are no primes: for example the sequence

$$n! + 2, \quad n! + 3, \quad ..., \quad n! + n .$$

Even Euler was discouraged by these tables, and thought that the distribution of prime numbers was "a mystery which the human mind will never be able to penetrate". At the end of the eighteenth century, however, Legendre and Gauss, independently, had the idea that *on average* the prime numbers obey simple laws. Gauss considered all the primes between a number x and the number $x + 1000$. If $N(x)$ is this number, he observed from the tables that, when x is large, the ratio $N(x)/1000$ is close to $1/\log x$. If we were considering a continuous layer of matter distributed over a straight line and of density $1/\log x$, the quantity contained between 2 and x would be

$$\mathrm{li}(x) = \int_2^x \frac{dt}{\log t} ,$$

which is called the *integral logarithm*. Gauss therefore thought that the number $\pi(x)$ of prime numbers contained between 2 and x must be "approximated by" the number $\mathrm{li}(x)$.

This idea was made precise by conjecturing that when x increases indefinitely, *the ratio* $\pi(x)/\mathrm{li}(x)$ *tends to* 1. This is what is called "the prime number theorem": if $p_1, p_2, ..., p_n, ...$ is the increasing sequence of prime numbers, it is shown that the theorem is equivalent to saying that *the ratio* $p_n/n \log n$ *tends to* 1.

An examination of the tables shows that these conjectures are plausible: for example, for $x = 4 \cdot 10^{16}$, we have

$$\pi(x) = 1\,075\,292\,778\,753\,150 , \qquad \mathrm{li}(x) - \pi(x) = 5\,538\,861 .$$

During the whole of the nineteenth century, many mathematicians set themselves to prove the prime number theorem; but it was not proved until 1896, almost simultaneously by J. Hadamard and C. de la Vallée-Poussin. Unfortunately, all the methods of proof are based on the study of a function denoted by $\zeta(s)$, introduced by Riemann, and which requires advanced techniques of analysis, which we cannot begin to describe.

Mathematicians were not, however, content with this success. They wanted to know the behaviour of the difference

$$\pi(x) - \mathrm{li}(x).$$

A conjecture of Riemann on the properties of his function $\zeta(s)$ would, if it were true, entail the result that

---

[5] In 1985, we knew $3\,424\,506$ numbers $p \le 10^9$ such that p and $p+2$ are both primes.

$$\frac{\pi(x) - \text{li}(x)}{x^{\frac{1}{2}+\alpha}}$$

tends to 0 for each exponent $\alpha > 0$. Unfortunately, despite 130 years of endeavour, nobody has yet been able to prove or disprove Riemann's hypothesis. This hypothesis remains one of the most important open problems of mathematics because its solution would bring about great advances in many areas of number theory.

For a long time it was believed that we always have $\pi(x) < \text{li}(x)$: the tables show that this is true for $x \leq 10^8$. Littlewood, however, established that there is an infinity of natural numbers x such that

$$\pi(x) - \text{li}(x) > \frac{\sqrt{x}}{2\log x},$$

and also an infinity of natural numbers x such that

$$\pi(x) - \text{li}(x) < -\frac{\sqrt{x}}{2\log x}.$$

It is still not known what is the smallest number x_0 for which $\pi(x) - \text{li}(x)$ changes sign, but it is certainly very large. This result obviously confirms the impression of extreme irregularity in the local distribution of primes.

In 1785, Legendre, in view of applications to the theory of quadratic forms, needed a refinement of Euclid's theorem, namely, to find out whether, if a and b are two natural numbers with no common prime factor (i.e, a and b are *coprime*) *there exists an infinity of primes in the arithmetical progression* $an + b$. In some particular cases this can easily be ascertained by generalizing Euclid's method in an appropriate way. For example, this can be done for the arithmetical progressions $4n + 3$ and $6n + 5$ (Appendix 1). But no one has ever succeeded in proving the general theorem by elementary methods: the proof was given by Dirichlet in 1837 by using functions which generalize Riemann's function $\zeta(s)$.

We denote by $\pi(x; a, b)$ the number of primes at most equal to x in the arithmetical progression $an + b$. Hadamard and de la Vallée Poussin extended their methods to obtain an estimate of $\pi(x; a, b)$, and showed that the ratio

$$\frac{\pi(x; a, b)}{x/\log x}$$

has the limit $1/\varphi(a)$, where $\varphi(a)$ is the number of primes m up to a and such that $0 < m < a$. This formula had already been conjectured by Legendre.

C. The Beginnings of Algebraic Geometry

We have seen (Chapter III, §8) that immediately after the invention of the method of coordinates, it appeared that the plane curves whose equation $P(x, y) = 0$ is given by a polynomial P of degree two are *conics*. This led, during the eighteenth and nineteenth centuries, to the classification of the curves of equation $P(x, y) = 0$ according to the *degree* of the polynomial P. Newton, for example, described the various forms of curves of degree three (called "cubics"). In the same way, surfaces of equation $P(x, y, z) = 0$, where P is a polynomial, were classified according to their degrees. The surfaces of degree two are called *quadrics*: they include in particular the sphere and certain surfaces of revolution studied by Archimedes.

During the nineteenth century, many mathematicians were enthralled by the study of families of conics or of quadrics satisfying certain conditions, and they discovered very elegant theorems in this field. The geometrical study of curves and surfaces of low degree also met with some pleasing successes. To cite only one example, it can be proved that on a cubic surface, either there are infinitely many straight lines – in which case we say that the surface is *ruled* – , or else there are *at most* 27 of them, which have remarkable geometric relations with one another.

These studies were the prelude to the opening up, in Riemann's work, of modern algebraic geometry, one of the areas of mathematics expanding most rapidly in our day after more than a hundred years of research in which analysis, algebra and topology all made substantial contributions. The highly technical nature of these subjects unfortunately makes it impossible for us to deal with them.

IV Appendix

1. Prime Numbers of the Form 4k-1 or 6k-1

A prime number other than 2 necessarily has one of the forms $4k+1$ or $4k-1$ with $k \geq 1$. The argument of Euclid, slightly modified, shows that there are *infinitely many* prime numbers of the form $4k - 1$.

We consider the increasing sequence

$$(1) \qquad p_1 = 3 < p_2 = 7 < \ldots < p_r$$

of prime numbers of the form $4k-1$, supposed to contain *all* the prime numbers of this form which are $\leq p_r$. We shall show that there is a prime number of the form $4k - 1$ which is not in this sequence, and so is $> p_r$.

We form the number

$$(2) \qquad N = 4p_1p_2 \ldots p_r - 1 .$$

This number is clearly not divisible by any of the numbers in the sequence (1). On the other hand, it cannot be a product of powers of prime numbers of the form $4k + 1$, because each product

$$(4a + 1)(4b + 1) = 4(4ab + a + b) + 1$$

is again of the form $4c+1$. Thus no product of prime factors of the form $4k+1$ can be equal to the number N defined by (2). Hence one of the prime factors of N is of the form $p = 4k - 1$, and p is different from each number of the sequence (1).

Each prime number other than 2 or 3 is necessarily of one of the forms $6k + 1$ or $6k - 1$. By replacing 4 by 6 in the above argument, we show that there are infinitely many prime numbers of the form $6k - 1$.

2. The Decomposition of $\zeta(s)$ as a Eulerian Product

Let

$$p_1 < p_2 < \ldots < p_r < \ldots$$

be the (*infinite*) increasing sequence of prime numbers. Also let s be an exponent *bigger than* 1. Euler's formula for prime numbers is the following:

(1) $(1 - \frac{1}{p_1^s})^{-1}(1 - \frac{1}{p_2^s})^{-1} \ldots (1 - \frac{1}{p_r^s})^{-1} \ldots =$

$$= 1 + \frac{1}{2^s} + \frac{1}{3^s} + \ldots + \frac{1}{n^s} + \ldots ,$$

where on the right-hand side every integer occurs once. To give a meaning to this formula, it is necessary to prove three things:

I) If, for each whole number m, we set

$$F(m) = 1 + \frac{1}{2^s} + \frac{1}{3^s} + \ldots + \frac{1}{m^s} ,$$

then the sequence $(F(m))$ has a *limit S*.

II) If we set

$$G(r) = (1 - \frac{1}{p_1^s})^{-1}(1 - \frac{1}{p_2^s})^{-1} \ldots (1 - \frac{1}{p_r^s})^{-1},$$

then the sequence $(G(r))$ has a *limit P*.

III) We have $P = S$.

Proof of I). If $m < m'$, then clearly $F(m) < F(m')$. As in §2, B, we consider the difference $F(2^N) - F(2^M)$ for $M < N$, which is written

$$(\frac{1}{(2^M + 1)^s} + \ldots + (\frac{1}{(2^{(M+1)s}}) + \ldots + (\frac{1}{(2^{N-1} + 1)^s} + \ldots + \frac{1}{2^{Ns}}).$$

In each parenthesis

$$\frac{1}{(2^k + 1)^s} + \frac{1}{(2^k + 2)^s} + \ldots + \frac{1}{2^{(k+1)s}} ,$$

there are 2^k terms, and each is less than $1/2^{ks}$, and so the sum of the terms in the parenthesis is less than $1/(2^{s-1})^k$. It follows that $F(2^N) - F(2^M)$ is smaller than the sum of the geometrical progression with first term $1/(2^{s-1})^M$ and with ratio $1/2^{s-1}$, so that

(2) $$F(2^N) - F(2^M) \leq \frac{a}{(2^{s-1})^M} - \frac{a}{(2^{s-1})^N} ,$$

where

$$a = (1 - \frac{1}{2^{s-1}})^{-1}.$$

Thus we have formed the intervals

$$\left[F(2^M), \ F(2^M) + \frac{a}{(2^{s-1})^M} \right] ,$$

which are nested, with length tending to 0, and which contain each number $F(m)$ for $m > 2^M$. Cauchy's principle shows that the limit S of the sequence $(F(m))$ exists as the unique number which is contained in all these intervals, and necessarily $F(m) < S$ for *each* natural number m.

Proof of II) *and* III). If $r < r'$, we have $G(r) < G(r')$. For each natural number N, there is a natural number r_N such that, for each natural number $n \leq 2^N$, the decomposition of n into prime factors contains only the prime numbers $p_1, p_2, \ldots, p_{r_N}$. The argument of §2, B) shows that we have $F(2^N) \leq G(r_N)$. On the other hand, for each *fixed* natural number r, $G(r)$ is the limit of

$$\frac{(1 - \frac{1}{p_1^{sM}})(1 - \frac{1}{p_2^{sM}}) \ldots (1 - \frac{1}{p_r^{sM}})}{(1 - \frac{1}{p_1^{s}})(1 - \frac{1}{p_2^{s}}) \ldots (1 - \frac{1}{p_r^{s}})}$$

as M increases indefinitely, and this number is at most $F((p_1 p_2 \ldots p_r)^M) < S$. For each $r > r_N$, all the numbers $G(r)$ are thus in the interval

$$[F(2^N),\ S]\,,$$

and Cauchy's principle shows that the sequence $(G(r))$ has a limit P equal to S.

The number S which is the common value of the two sides of (1), is denoted by $\zeta(s)$.

3. Lagrange's Method for the Solution of $ax^2 + bxy + cy^2 = n$ in Integers

Some preliminary propositions and remarks are necessary.

I) *Bezout's theorem.* If a and b are two natural numbers, there exist two (positive or negative) integers m, n such that

$$(1) \qquad\qquad ma + nb = d\,,$$

where d is the greatest common divisor of m and n.

This can be seen from successive Euclidean divisions: if $a \geq b$, we carry out successively the divisions

$$(2) \quad \begin{cases} a = bq + r_1\,, & \text{where} \quad r_1 < b, \\ b = r_1 q_1 + r_2\,, & \text{where} \quad r_2 < r_1, \\ r_1 = r_2 q_2 + r_3\,, & \text{where} \quad r_3 < r_2, \\ \cdots\cdots\cdots\cdots\cdots\cdots\cdots\cdots\cdots, \end{cases}$$

and, since the remainders r_j are decreasing, we necessarily reach a zero remainder

$$r_{k-1} = r_k q_k + r_{k+1}\,,$$
$$r_k = r_{k+1} q_{k+1}\,.$$

Since each common factor of a and b necessarily divides each of the r_j, we have $d = r_{k+1}$. But if we solve successively each equation (2), we obtain $r_1 = a - bq$,

$r_2 = b - r_1 q_1, \ldots$, and we see that all the r_j are of the form $ua + vb$, where u and v are positive or negative integers.

II) To solve

$$(3) \qquad\qquad ax^2 + bxy + cy^2 = n \,,$$

we can limit ourselves to the case where the greatest common divisor of a, b and c is 1. For, if d is this common factor, then d must divide n, and, if $n = dn_1$, $a = da_1$, $b = db_1$, and $c = dc_1$, then equation (3) is equivalent to

$$a_1 x^2 + b_1 xy + c_1 y^2 = n_1 \,.$$

III) We can limit ourselves to seeking solutions of (3) such that the greatest common divisor of x and y is 1 (the so-called *primitive solutions*). In fact, if $x = dx_1$ and $y = dy_1$, then d^2 must divide n, and, setting $n = d^2 n_1$, we have

$$a x_1^2 + b x_1 y_1 + c y_1^2 = n_1 \,.$$

IV) It can be supposed that the discriminant $D = b^2 - 4ac$ is not a square. For otherwise the equation $at^2 + bt + c = 0$ has two *rational* roots $t_1 = v_1/u_1$ and $t_2 = v_2/u_2$, written in the form of irreducible fractions (with u_1, u_2 positive and v_1, v_2 either positive or negative). Then, since

$$at^2 + bt + c = a(t - t_1)(t - t_2) \,,$$

equation (3) can be written

$$a(u_1 x - v_1 y)(u_2 x - v_2 y) = n u_1 u_2 \,.$$

To solve this, we decompose $n u_1 u_2 / a$ in all possible ways as a product of two factors r, s, and then solve the system

$$u_1 x - v_1 y = r \,, \qquad u_2 x - v_2 y = s \,,$$

keeping only the integral solutions (if there are any).

V) If b is even, then the discriminant D is a multiple of 4; if b is odd, then $D - 1$ is a multiple of 4. Set $\rho = 0$ if b is even and $\rho = 1$ if b is odd, so that $(D - \rho)/4$ is always an integer.

VI) We write $a \equiv b \pmod{m}$, and we say that "a is congruent to b modulo m" as an abbreviation of the property

$$a - b \qquad \text{is a multiple of} \qquad m \,.$$

Lagrange's theorem is then the following:

For there to be a quadratic form $ax^2 + bxy + cy^2$ with discriminant D such that equation (3) has a primitive solution, it is necessary and sufficient that the congruence

$$z^2 + \rho z - \frac{D - \rho}{4} \equiv 0 \pmod{n}$$

has a solution z such that $0 \leq z < n$.

Necessity. Let us suppose that (α, γ) is a primitive solution of (3). By Bezout's theorem, there exist β, δ such that $\alpha\delta - \beta\gamma = 1$. If we make the change of variables

$$x = \alpha X + \beta Y$$
$$y = \gamma X + \delta Y,$$

then we obtain an equation

$$nX^2 + (2r + \rho)XY + sY^2 = n,$$

which has the solution $(1, 0)$ corresponding to (α, γ). Since the discriminant has not changed, we have

$$(5) \qquad\qquad (2r + \rho)^2 - 4sn = D,$$

and, since $\rho^2 = \rho$, this can be written as

$$(6) \qquad\qquad r^2 + \rho r - \frac{D - \rho}{4} = sn.$$

In other words, r is a solution of the congruence (4), and so the same is true of the remainder z when r is divided by n.

Sufficiency. If there exists a solution r of the congruence (4), this amounts to saying that there is an integer s (positive or negative) satisfying (6), and hence (5). But this means that the quadratic form

$$nx^2 + (2r + \rho)xy + sy^2$$

has D as its discriminant, and it takes the value n when $x = 1$, $y = 0$.

Let us take for example the equation

$$(7) \qquad\qquad x^2 + y^2 = n.$$

It can be shown that there is only one class of quadratic forms with discriminant $D = -4$. Lagrange's theorem thus shows that the necessary and sufficient condition for equation (7) to have solutions in the natural numbers is that the congruence

$$(8) \qquad\qquad z^2 \equiv -1 \pmod{n}$$

has a solution.

Let us suppose that n is an *odd prime*. We apply a famous theorem of Fermat, which we shall come across again in Chapter V, Appendix 3, namely, that for each integer $a \neq 0$, we have

$$(9) \qquad\qquad a^{n-1} \equiv 1 \pmod{n}.$$

If a congruence

$$z^2 \equiv a \pmod{n}$$

has a solution, it must be the case that

$$a^{(n-1)/2} \equiv z^{2(n-1)/2} \equiv 1 \pmod{n},$$

and it can be seen (*loc. cit.*) that this condition is sufficient. Since $(-1)^{(n-1)/2}$ is congruent to -1 if $n = 4k + 3$ and to 1 if $n = 4k + 1$, we see that each odd prime number of the form $4k + 1$ is the sum of two squares, the theorem stated by Fermat and proved by Euler.

4. Bernoulli Numbers and the Zeta Function

This appendix is intended for readers who have a certain familiarity with ordinary power series expansions.

We have seen in Chapter III, §9 and Appendix 4 that, in his original calculations of primitives, Fermat, following Archimedes, had had to evaluate sums

$$(1) \qquad S_m(n) = 1^m + 2^m + \ldots + n^m$$

for an arbitrary natural number $m \geq 1$. It is sufficient for this to find a polynomial $S_m(t)$ satisfying the equation

$$(2) \qquad S_m(t+1) - S_m(t) = t^m.$$

It is immediately checked from the binomial formula that, if $S_m(t)$ is a polynomial of degree $m + 1$, its coefficients are completely determined by equation (2), *except for* the constant term. This latter term is given by the condition that $S_m(1) = 1$, which follows from (1).

Jacob Bernoulli gave an elegant description of the polynomials $S_m(t)$, which can be presented in the following fashion. Slightly modifying equation (2), we consider for each $n \geq 1$ polynomials $\varphi_n(x)$ of degree n which satisfy the equation

$$(3) \qquad \varphi_n(x+1) - \varphi_n(x) = nx^{n-1}.$$

We consider the "generating function"

$$(4) \qquad F(z, x) = \sum_{n=1}^{\infty} \frac{\varphi_n(x)}{n!} z^{n-1}$$

of this sequence, where the series must be convergent for z in a neighbourhood of 0 and for any x. It follows from (3) that we must have

$$(5) \qquad F(z, x+1) - F(z, x) = \sum_{n=1}^{\infty} \frac{x^{n-1} z^{n-1}}{(n-1)!} = e^{zx}.$$

If we look for a solution of (5) of the form

$$F(z,x) = e^{zx}f(z) + g(z),$$

then we obtain $f(z) = 1/(e^z - 1)$, $g(z)$ being arbitrary. But $f(z)$ is not defined for $z = 0$, and this singularity is eliminated by setting $g(z) = -1/z$. Thus the *Bernoulli polynomials* $\varphi_n(x)$ are defined by the equation

(6) $$\frac{e^{zx}}{e^z - 1} = \frac{1}{z} + \sum_{n=1}^{\infty} \frac{\varphi_n(x)}{n!} z^{n-1}.$$

For $x = 0$, we are led to the series expansion

(7) $$\frac{1}{e^z - 1} = \frac{1}{z(1 - u(z))} = \frac{1}{z}\left(1 + u(z) + u(z)^2 + \ldots + u(z)^n + \ldots\right)$$

with

(8) $$u(z) = \frac{1 + z - e^z}{z} = -\frac{z}{2!} - \frac{z^2}{3!} - \ldots - \frac{z^n}{(n+1)!} - \ldots.$$

Since

$$\frac{1}{e^z - 1} = -\frac{1}{2} + \frac{e^z + 1}{2(e^z - 1)}$$

and since

$$\frac{e^{-z} + 1}{e^{-z} - 1} = -\frac{e^z + 1}{e^z - 1},$$

the series expansion of

$$\frac{1}{e^z - 1} + \frac{1}{2}$$

contains only terms of *odd* exponent. If we write

(9) $$\frac{1}{e^z - 1} = \frac{1}{z} - \frac{1}{2} + \sum_{n=1}^{\infty} (-1)^{n-1} \frac{B_n}{(2n)!} z^{2n-1},$$

it follows from (7) and (8) that the numbers B_n are *rational*. Calculations give the values of the first numbers as follows:

$$B_1 = \frac{1}{6}, \quad B_2 = \frac{1}{30}, \quad B_3 = \frac{1}{42}, \quad B_4 = \frac{1}{30}, \quad B_5 = \frac{5}{66}, \quad B_6 = \frac{691}{2730},$$

$$B_7 = \frac{7}{6}, \quad B_8 = \frac{3617}{510}, \quad B_9 = \frac{43867}{798}, \quad B_{10} = \frac{174611}{330}, \quad B_{11} = \frac{854513}{138},$$

$$B_{12} = \frac{236364091}{2730}, \quad B_{13} = \frac{8553103}{6}, \quad B_{14} = \frac{23749461029}{870}.$$

These are called the *Bernoulli numbers*. They play an important and rather mysterious role in such different areas of mathematics as analysis, number theory (notably in the study of Fermat's equation $x^n + y^n = z^n$), and differential topology. If we multiply (8) and the power series expansion of e^{xz}, we arrive at the expression

$$(10) \quad \varphi_n(x) = x^n - \frac{n}{2}x^{n-1} + \binom{n}{2}B_1 x^{n-2} - \binom{n}{4}B_2 x^{n-4} + \binom{n}{6}B_3 x^{n-6} - \ldots$$

for the Bernoulli polynomials, where the sum stops at the last term where the exponent is ≥ 0. We have the formulae

$$(11) \qquad\qquad \varphi_n(1 - x) = (-1)^n \varphi_n(x)$$

$$(12) \qquad\qquad \varphi_{2k+1}(0) = 0, \quad \varphi_{2k}(0) = (-1)^{k+1} B_k$$

$$(13) \qquad\qquad \varphi_n'(x) = n\varphi_{n-1}(x) \quad \text{for} \quad n \geq 2.$$

These follow immediately from (6) and the equation

$$\frac{e^{z(1-x)}}{e^z - 1} = -\frac{e^{-xz}}{e^{-z} - 1}.$$

In Chapter V, Appendix 5, we shall show that, for $0 \leq x \leq 1$, the Bernoulli polynomials have the following remarkable expansions in trigonometric series (for $k \geq 1$):

$$(14) \qquad\qquad \varphi_{2k}(x) = (-1)^{k+1} 2\,(2k)! \sum_{n=1}^{\infty} \frac{\cos 2n\pi x}{(2n\pi)^{2k}},$$

$$(15) \qquad\qquad \varphi_{2k+1}(x) = (-1)^{k+1} 2\,(2k+1)! \sum_{n=1}^{\infty} \frac{\sin 2n\pi x}{(2n\pi)^{2k+1}}.$$

If we set $x = 0$ in (14), we find the formulae obtained by Euler

$$(16) \qquad \zeta(2k) = 1 + \frac{1}{2^{2k}} + \frac{1}{3^{2k}} + \ldots + \frac{1}{n^{2k}} + \ldots = \frac{2^{2k-1}}{(2k)!} B_k \pi^{2k}$$

for the values of $\zeta(2k)$, which were so much admired by his contemporaries. For low values of k, we thus have

$$\zeta(2) = \frac{\pi^2}{2 \cdot 3}, \quad \zeta(4) = \frac{\pi^4}{2 \cdot 3^2 \cdot 5}, \quad \zeta(6) = \frac{\pi^6}{3^3 \cdot 5 \cdot 7}, \quad \zeta(8) = \frac{\pi^8}{2 \cdot 3^3 \cdot 5^2 \cdot 7}.$$

It has been known for a long time that π and all its powers are *irrational* numbers, and so the same is true for the numbers $\zeta(2k)$. Since we do not have any expression for $\zeta(2k + 1)$ analogous to (16), the question has been asked since the time of Euler whether the numbers $\zeta(2k + 1)$ are themselves also irrational, and Euler himself had tried to prove that this was the case. For 200 years not the smallest advance was made towards solving this problem, which thus appeared to be intractable. However, in 1977, the French mathematician R. Apéry thought of a very ingenious procedure, which, although using only results known at the time of Euler, enabled him to show that $\zeta(3)$ *is irrational*. Unfortunately, despite much research, no one has yet succeeded in finding analogous procedures which could be applied to $\zeta(5), \zeta(7)$, etc.

Returning to the original problem of Fermat, we find from (3) that

$$\varphi_{m+1}(1) - \varphi_{m+1}(0) = 0$$
$$\varphi_{m+1}(2) - \varphi_{m+1}(1) = (m+1) \cdot 1^m$$
$$\varphi_{m+1}(3) - \varphi_{m+1}(2) = (m+1) \cdot 2^m$$

$$\dotfill$$

$$\varphi_{m+1}(n+1) - \varphi_{m+1}(n) = (m+1) \cdot n^m,$$

whence, adding the corresponding sides of these equalities, we obtain

$$S_m(n) = \frac{1}{m+1} \left(\varphi_{m+1}(n+1) - \varphi_{m+1}(0) \right),$$

and formula (10) gives an explicit expression for $S_m(n)$ in terms of the Bernoulli numbers.

V New Objects and New Methods

The eighteenth century had been a dazzling era for its intensive development of techniques which had been introduced in the seventeenth century, especially in analysis and its various applications; and this is equally true of applications to other mathematical disciplines such as geometry and probability theory as to mechanics and astronomy, whose success in the prediction of natural phenomena is well-known[1]. Strangely, however, the century seemed to end in an impasse. The great mathematicians of the middle of the century departed, Daniel Bernoulli in 1782, Euler and d'Alembert in 1783. Lagrange, at scarcely fifty years of age, judged that the era of progress in pure mathematics was over, and after 1785 he and Monge turned their attention to physics and chemistry, while Laplace was exclusively concerned with mechanics and probability theory. Then came the revolutionary years, which put men of science at the service of the nation and of the war, with the result that the decade 1786-1796 is not marked in France by any significant new mathematical result. Barren periods such as this, connected with social disturbances, have recurred in our time in several countries, but at that epoch, no country apart from France had any active mathematician comparable to those we have just named, which meant that the sterility was universal.

It was therefore a true renaissance which began with Gauss in 1796. Trained exclusively by reading the works of his predecessors, he was to renew the whole of mathematics in fifteen years. From the early years of the nineteenth century he was no longer alone. One of the most productive innovations of the Revolution was the creation of genuine higher education in science, purveyed by eminent teachers and open to all[2]. The Ecole Polytechnique – although designed mainly to train military men and engineers – was to be continuously for seventy-five years a nursery for mathematicians, physicists and chemists of the first order. Apart from Gauss, mathematicians trained at the Ecole Polytechnique were without rivals up to 1825.

It was the ideas formulated in the eighteenth century and even earlier which formed the basis for this renaissance. But almost at once the style and the content were changed, not only by the famous "return to rigour" inaugu-

---

[1] A classic example is the gyroscope, with its paradoxical movements which cannot be understood without using the equations of solid dynamics.

[2] Under the Ancien Régime only schools designed to train future officers gave courses in mathematics allowing a little space to the infinitesimal calculus, and these schools were scarcely accessible to the common people.

rated by Gauss, Bolzano, Cauchy and Abel (see Chapter VI, §2, A), but also by the introduction of new mathematical objects, which differ from classical objects because *they can no longer be represented by "pictures" accessible to our senses.*

The proliferation of new ideas in all the traditional areas of mathematics – arithmetic, algebra, geometry, analysis – continued uninterruptedly throughout the nineteenth century, which we see now as a transitional epoch, forming the bridge between "classical" mathematics and our own. It was a period of astonishing productivity: on the one hand there was the *discovery* of concepts which were themselves to become the bases of entirely new areas of mathematics – group theory, topology, function spaces, etc. – areas which in our time have acquired an amplitude comparable to those of the traditional areas mentioned above; on the other hand a greater *depth* was brought to old ideas, making it possible to appreciate better the true significance of the axioms.

Little by little there emerged a general idea which was to be given precision in the twentieth century, that of *structure* at the basis of a mathematical theory. It arises from the observation that the primary role in a theory is played by the *relations* between the mathematical objects concerned rather than by the *nature* of these objects; and thus it may be that in two very different theories relations are expressed *in the same way* in both. The system of these relations and of their consequences form one and the same structure "underlying" the two theories.

The reader will be shown by means of several accessible examples how there appeared all through the course of the nineteenth century a number of grand structures which are at the foundation of the mathematics of our day. It must be stressed that, in almost all cases, these appeared in response to a *need*, in order to attack successfully problems inherited from *classical* mathematics, and were not due to the fantasy of a mathematician, creating new abstract notions without any precise objective.

The fact that the same structure can appear in two very different theories was to bring an ever-increasing awareness of the fundamental *unity* of *all* mathematics, overriding the traditional partitioning based on the nature of the objects studied. But a long time was needed for this view of the subject's wholeness to emerge, and during the nineteenth century each new theory was most commonly developed without regard for its possible links with others. Only at the end of this chapter shall we be able to sketch this view of wholeness, the foundation of our mathematics and the motive force of its progress (see §5, A).

1. New Calculations

A. Complex Numbers

The earliest of the great mathematical discoveries that went clearly beyond the knowledge of classical times was the formula for the solution of an arbitrary cubic equation

$$(1) \qquad x^3 - px = q \quad \text{(with} \quad p > 0, \quad q > 0\text{)}.$$

We note, incidentally, that the arbitrary cubic equation

$$x^3 + a_1 x^2 + a_2 x + a_3 = 0$$

can be expressed in the above form, for if we set $x = y - a_1/3$, we find that the equation for y is written as

$$y^3 - \frac{a_1^2}{3} - a_2)y + \frac{2a_1^3}{27} - \frac{a_1 a_2}{3} + a_3 = 0.$$

The solution of (1) was probably obtained at the beginning of the sixteenth century by Scipione del Ferro, professor at the University of Bologna, although he did not publish it but only communicated it to several pupils and friends. His method was to look for a solution of the form $x = u + v$; we have

$$x^3 - px = u^3 + v^3 + 3uv(u + v) - p(u + v)$$
$$= u^3 + v^3 + (3uv - p)(u + v).$$

The term involving $u + v$ *disappears* in the case where $3uv - p = 0$, and as a consequence the number $x = u + v$ is a solution of (1) if, *simultaneously*, we have

$$(2) \qquad uv = p/3,$$
$$(3) \qquad u^3 + v^3 = q.$$

But from (2) we have

$$(4) \qquad u^3 v^3 = \frac{p^3}{27},$$

and the classical theory of quadratic equations shows that the relations (3) and (4) imply that u^3 and v^3 are the two roots of the equation

$$(5) \qquad X^2 - qX + \frac{p^3}{27} = 0,$$

that is to say

$$(6) \qquad u^3 = \frac{q}{2} + \sqrt{\frac{q^2}{4} - \frac{p^3}{27}}, \qquad v^3 = \frac{q}{2} - \sqrt{\frac{q^2}{4} - \frac{p^3}{27}},$$

whence finally

(7) $$x = u + v = \sqrt[3]{\frac{q}{2} + \sqrt{\frac{q^2}{4} - \frac{p^3}{27}}} + \sqrt[3]{\frac{q}{2} - \sqrt{\frac{q^2}{4} - \frac{p^3}{27}}}.$$

(This formula is generally known under the name "Cardan's formula" because Cardan was the first to make it known.)

Since the invention of the differential calculus, we have known how to study the variation of the function $y = x^3 - px - q$ by taking its derivative; it follows easily that equation (1) has

$$\begin{cases} \text{one root} & \text{if } \dfrac{q^2}{4} - \dfrac{p^3}{27} > 0 \\[2mm] \text{three roots} & \text{if } \dfrac{q^2}{4} - \dfrac{p^3}{27} < 0. \end{cases}$$

But, in the second case, formula (7) no longer seems to make sense: in fact, by the rule of signs, every square of a real number (positive or negative) is positive, and so a negative number cannot have a square root. However, confronted by these difficulties, the Italian algebraists and their successors did not hesitate to perform calculations with these numbers $\sqrt{-a}$, where $a > 0$, *as if* they existed. That is to say, they applied the usual algebraic rules to them together with the rule $(\sqrt{-a})^2 = -a$. For example, R. Bombelli applied Cardan's formula to the equation

$$x^3 - 15x = 4,$$

which he knew to have a real root $x = 4$. Since

$$x^3 - 15x - 4 = (x - 4)(x^2 + 4x + 1),$$

there are two other real roots, $-2 \pm \sqrt{3}$. The root given by Cardan's formula is written

(8) $$x = \sqrt[3]{2 + \sqrt{-121}} + \sqrt[3]{2 - \sqrt{-121}}.$$

Bombelli applied the binomial expansion

$$(a + b)^3 = a^3 + 3a^2b + 3ab^2 + b^3 :$$

$$(2 + \sqrt{-1})^3 = 2^3 + 3 \cdot 2^2\sqrt{-1} + 3 \cdot 2(\sqrt{-1})^2 + (\sqrt{-1})^3$$

and, replacing $(\sqrt{-1})^2$ by -1 and $(\sqrt{-1})^3$ by $-1 \cdot \sqrt{-1}$, obtained

$$(2 + \sqrt{-1})^3 = 2 + 11\sqrt{-1} = 2 + \sqrt{(-1)(11)^2} = 2 + \sqrt{-121}.$$

In the same way, he calculated

$$(2 - \sqrt{-1})^3 = 2 - \sqrt{-121},$$

and obtained from (8) that

$$x = 2 + \sqrt{-1} + 2 - \sqrt{-1} = 4.$$

This result, he said, seemed to him to be satisfying even if "sophisticated"! His predecessor Cardan, who was probably the first to dare to do such calculations, spoke of "moral torture" in referring to them! This did not prevent their successors from pursuing these calculations ever more intensively for over 200 years with numbers which they deemed to be "impossible" or "imaginary". If $b > 0$, we have $-b = b \times (-1)$, and by analogy with the formula $\sqrt{bb'} = \sqrt{b} \cdot \sqrt{b'}$ for positive numbers b and b' , we write $\sqrt{-b} = \sqrt{b} \times \sqrt{-1}$. Thus all imaginary numbers are reduced to the form $x + y\sqrt{-1}$, where x and y are real numbers, or $x + iy$, using the abbreviation i for $\sqrt{-1}$ introduced by Euler. This persistance in using meaningless expressions was due to the fact that algebraists found in these calculations the possibility of obtaining *general* theorems, without the need to distinguish several cases : a quadratic equation *always* has either one double root or two simple roots, real *or imaginary* . A cubic equation with real coefficients always has at least one real root,[3] and is therefore written as

$$(x - a)(x^2 + px + q) = 0$$

and the sum of the multiplicities of the roots is *always* three if we count imaginary roots. By the beginning of the seventeenth century, A. Girard had arrived at the conviction that an equation of degree n

$$a_0 x^n + a_1 x^{n-1} + \ldots + a_n = 0, \quad \text{with} \quad a_0 \neq 0,$$

has real or imaginary roots the sum of whose multiplicities is *always* n. This is what is called the "fundamental theorem of algebra". But it was not until the end of the eighteenth century that it was justified and fully proved by Gauss.

By that time, Gauss, as well as two less well-known mathematicians, Wessel and Argand, had at last independently succeeded in giving a meaning to

---

[3] The polynomial $x^3 + a_1 x^2 + a_2 x + a_3$ is written for $x \neq 0$ as

(1) $$x^3 1 + \frac{a_1}{x} + \frac{a_2}{x^2} + \frac{a_3}{x^3}),$$

and, if $| x | > 1$, then we have

$$\left| \frac{a_1}{x} + \frac{a_2}{x^2} + \frac{a_3}{x^3} \right| \leq \frac{|a_1| + |a_2| + |a_3|}{| x |}.$$

Hence

$$1 + \frac{a_1}{x} + \frac{a_2}{x^2} + \frac{a_3}{x^3} \geq 1 - \frac{|a_1| + |a_2| + |a_3|}{x},$$

and, if $|x| \geq 1$ and $|x| > |a_1| + |a_2| + |a_3| = A$, then the product (1) has the same sign as x^3, that is to say, it is < 0 if $x < -A - 1$ and > 0 if $x > A + 1$. Stevin's method (Chapter III, §6) proves the existence of a real root a of the equation

$$x^3 + a_1 x^2 + a_2 x + a_3 = 0.$$

calculations with the numbers $a + bi$, which following Gauss are called *"complex numbers"*. It is sufficient to identify $a + bi$ with a *point* with coordinates (a, b) with respect to rectangular axes in a plane. An equivalent interpretation, proposed later by Hamilton, is to consider $a + bi$ as a *pair*

$$\begin{pmatrix} a \\ b \end{pmatrix}$$

of real numbers, with the rules of calculation

(9) $$\begin{pmatrix} a \\ b \end{pmatrix} + \begin{pmatrix} a' \\ b' \end{pmatrix} = \begin{pmatrix} a + a' \\ b + b' \end{pmatrix}$$ for addition

(10) $$\begin{pmatrix} a \\ b \end{pmatrix} \times \begin{pmatrix} a' \\ b' \end{pmatrix} = \begin{pmatrix} aa' - bb' \\ ab' + ba' \end{pmatrix}$$ for multiplication.

The pair $\begin{pmatrix} a \\ 0 \end{pmatrix}$ is identified with the real number a,[4] in such a way that the real numbers appear as special complex numbers, and the rule (10) gives in particular

$$\begin{pmatrix} 0 \\ 1 \end{pmatrix}^2 = \begin{pmatrix} -1 \\ 0 \end{pmatrix} = -1;$$

that is to say the "impossible" calculation $i^2 = -1$ if

$$i = \begin{pmatrix} 0 \\ 1 \end{pmatrix} .$$

From the eighteenth century on, mathematicians had also used complex numbers in analysis and even in number theory without justifying their use, so obtaining correct results about "true" numbers, which appeared to be miraculous. Once the calculations with complex numbers were justified, these bold forays became the germs of enormously extensive theories which were to be developed all through the nineteenth century: the theory of analytic functions of complex variables by Cauchy, Riemann, Weierstrass and Poincaré, and the theory of algebraic numbers from Gauss and Dirichlet to Dedekind and Hilbert. Unfortunately we can only mention these theories in our general list (§5, A) because of the technical knowledge needed to grapple with their study.

B. Vectors

The notion of addition defined in (9) had already arisen much earlier in other contexts. If A is the point $\begin{pmatrix} a \\ b \end{pmatrix}$ and A' the point $\begin{pmatrix} a' \\ b' \end{pmatrix}$, then the point

---

[4] This identification is legitimate, since we have

$$\begin{pmatrix} a \\ 0 \end{pmatrix} + \begin{pmatrix} b \\ 0 \end{pmatrix} = \begin{pmatrix} a + b \\ 0 \end{pmatrix} \quad \text{and} \quad \begin{pmatrix} a \\ 0 \end{pmatrix} \times \begin{pmatrix} b \\ 0 \end{pmatrix} = \begin{pmatrix} ab \\ 0 \end{pmatrix} .$$

$$\begin{pmatrix} a + a' \\ b + b' \end{pmatrix}$$

is the endpoint C of the diagonal of the parallelogram of which OA and OB are two sides[5] (Figure 35). Now from the beginning of the seventeenth century (and doubtless even earlier, at least in special cases), this "parallelogram law" was used to determine the speed of a point propelled by a movement which is a "compound" of two others (an example is the description in kinetic terms of Archimedes' spiral, cf. Chapter III, §8). Speed, according to this conception, is represented by a *"vector"*: that is, a segment of line going in the direction of the movement which follows the tangent to the trajectory, its length being in proportion to the measure of its speed. Roberval in particular made a cunning use of this rule to determine the tangents to certain curves. Later it was by this same construction that Newton defined the composition of forces applied to a single point by representing them as segments of line originating from this point. The properties of this "addition" are altogether analogous to those of the addition of real numbers.

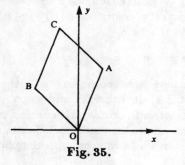

Fig. 35.

Of course, one might equally well carry out this parallelogram construction in space, which amounts to considering *triples*

$$\begin{pmatrix} a \\ b \\ c \end{pmatrix}$$

of real numbers and to defining their addition by

(11)
$$\begin{pmatrix} a \\ b \\ c \end{pmatrix} + \begin{pmatrix} a' \\ b' \\ c' \end{pmatrix} = \begin{pmatrix} a + a' \\ b + b' \\ c + c' \end{pmatrix}.$$

But why stop at triples? There is no reason why a system of n numbers should not be called *an n-dimensional vector*

---

[5] When O, A and B are aligned, $C = A + B$ for the addition defined geometrically in Chapter III, §5.

$$(12) \qquad \mathbf{x} = \begin{pmatrix} a_1 \\ a_2 \\ \vdots \\ a_n \end{pmatrix}$$

for an arbitrary natural number n, even if, when $n > 3$, a segment of a line can no longer be associated with it. The numbers a_j $(1 \leq j \leq n)$ are the *components* of the vector $\mathbf{x}$ and the sum $\mathbf{x} + \mathbf{x}'$ of two vectors is the vector whose components are the sums $a_j + a'_j$ of the components of $\mathbf{x}$ and of $\mathbf{x}'$.

Another operation on vectors arises from the notion of *similarity* in the plane (see §E, below): this is the "product" of a vector $\mathbf{x} = \begin{pmatrix} a \\ b \end{pmatrix}$ by a number α (called a *scalar*, to distinguish it from vectors), defined by

$$\alpha \cdot \mathbf{x} = \begin{pmatrix} \alpha a \\ \alpha b \end{pmatrix}.$$

This can immediately be generalized; for a vector $\mathbf{x}$ equal to a system (12) of n numbers, $\alpha \cdot \mathbf{x}$ is the vector having αa_j $(1 \leq j \leq n)$ as components. This allows one to express $\mathbf{x}$ in the form

$$\mathbf{x} = a_1 \mathbf{e}_1 + a_2 \mathbf{e}_2 + \ldots + a_n \mathbf{e}_n,$$

where $\mathbf{e}_1$ is the vector whose *first* component is 1, the others 0, $\mathbf{e}_2$ the one whose *second* component is 1, the others 0, and so on.

In the middle of the nineteenth century, Grassmann introduced yet another operation on vectors which allows the use of coordinates to be much simplified. This is the *scalar product* $\mathbf{u}.\mathbf{v}$ of two vectors in the plane

$$\mathbf{u} = \begin{pmatrix} u_1 \\ u_2 \end{pmatrix}, \qquad \mathbf{v} = \begin{pmatrix} v_1 \\ v_2 \end{pmatrix},$$

which is by definition the *number*

$$u_1 v_1 + u_2 v_2.$$

At the same time, this operation makes it possible to write the *length* of the vector $\mathbf{u}$, which is denoted by $\|\mathbf{u}\|$ and which is given by

$$(13) \qquad \|\mathbf{u}\| = \sqrt{\mathbf{u} \cdot \mathbf{u}},$$

and the cosine of the angle θ between $\mathbf{u}$ and $\mathbf{v}$, given by

$$(14) \qquad \cos \theta = \frac{\mathbf{u} \cdot \mathbf{v}}{\|\mathbf{u}\|.\|\mathbf{v}\|},$$

in such a way that the condition of *orthogonality* of two vectors is written simply as

$$(15) \qquad \mathbf{u} \cdot \mathbf{v} = 0.$$

In addition, the scalar product obeys very simple rules of calculation:

(16) $$\mathbf{u} \cdot \mathbf{v} = \mathbf{v} \cdot \mathbf{u}$$
(17) $$(\mathbf{u} + \mathbf{u}') \cdot \mathbf{v} = \mathbf{u} \cdot \mathbf{v} + \mathbf{u}' \cdot \mathbf{v}$$
(18) $$(\alpha \mathbf{u}) \cdot \mathbf{v} = \alpha(\mathbf{u} \cdot \mathbf{v})$$

for each scalar α.

These properties make many of the propositions of Euclid almost obvious, once they are translated into the language of vectors and scalar products. For example, the length a of the hypotenuse of a right-angled triangle of which b and c are the lengths of the sides adjacent to the right angle is the same as that of the vector $\mathbf{u} + \mathbf{v}$, where $\mathbf{u}$ and $\mathbf{v}$ are two perpendicular vectors of lengths b and c (Figure 36). We have, by (13), (16) and (17), that

$$a^2 = \|\mathbf{u} + \mathbf{v}\|^2 = (\mathbf{u} + \mathbf{v}).(\mathbf{u} + \mathbf{v}) = \mathbf{u} \cdot \mathbf{u} + 2\,\mathbf{u} \cdot \mathbf{v} + \mathbf{v} \cdot \mathbf{v},$$

and since, by hypothesis $\mathbf{u} \cdot \mathbf{v} = 0$, we have

$$\|\mathbf{u} + \mathbf{v}\|^2 = \|\mathbf{u}\|^2 + \|\mathbf{v}\|^2,$$

whence $a^2 = b^2 + c^2$. This is exactly Pythagoras's theorem.

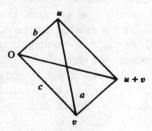

Fig. 36.

The notation $\mathbf{u} \cdot \mathbf{v}$ and the preceding properties extend without modification to vectors in space if we set

$$\mathbf{u} \cdot \mathbf{v} = u_1 v_1 + u_2 v_2 + u_3 v_3,$$

where u_1, u_2, u_3 are the three components of $\mathbf{u}$, and v_1, v_2, v_3 those of $\mathbf{v}$. Thus in the utilization of the scalar product, we not only dispense with a system of coordinates, but even with the dimension. This was to be recalled when, in the twentieth century, the space of atomic physics had become "Hilbert space" (Appendix 4, §C).

C. Algebraic Calculation on Functions

We have seen (Chapter III, §8) that the idea of a *function* of a real variable taking real values, which emerged at the end of the seventeenth century, could, by the expedient of *graphing* it, be considered as deduced by "abstraction" from a "rough" notion of plane geometry accessible to the senses. But from the beginning of the eighteenth century, it became necessary to consider functions (taking real values) of *several* real variables $x, y, z, \ldots$, for example

$$f(x, y, z) = xy + yz + zx.$$

For two variables, it is still possible to speak of the "graph", which is this time a *surface* in space plotted with respect to three coordinate axes; for example, the function $f(x, y) = xy$ has as its graph the surface with equation $z - xy = 0$, a quadric called the "hyperbolic paraboloid". But as soon as the number of variables exceeds two, we must abandon any "picture". In the eighteenth century it was merely said that $f(x, y, z, \ldots)$ is the value obtained by a certain number of "operations" carried out on the values given to the variables $x, y, z, \ldots$. There was moreover never a very precise definition of what was meant by "operation"; of course, the algebraic operations of sum, product, and quotient were always mentioned, but often there were added to them exponentiation x^y, logarithms, trigonometric functions, and the roots of equations (even if not resolved), etc. Little by little the notion was extended, and Dirichlet finally gave a general definition: all that must be assumed is that, for each system of values of the variables, there is *some* procedure associating with this system a *well-determined* real number. Dirichlet gives an example which is still famous: the function is defined by the condition that $f(x) = 0$ if x is a rational number and $f(x) = 1$ if x is an irrational number. It is obvious that no "graph" can be drawn of this function.

Since John Bernoulli, the notation φx or $f(x)$ has been used for the value of an unspecified function at the value x of the variable. Since the beginning of the nineteenth century, however, it has become customary (at least for the functions of one variable) to think of a function as *a single object*, not a succession of values, and to write simply f; for example, sin or log. *Calculations* are done on these objects: obviously $f + g$ and fg are defined by

$$(f + g)(x) = f(x) + g(x), \qquad (fg)(x) = f(x)g(x).$$

But there are other operations on the functions: the *composition* of two functions f and g, which makes $f(g(x))$ correspond to x, and which in our day we write $f \circ g$ to distinguish it from the product fg[6]; the *inverse* (or "reciprocal") f^{-1} which, when it means anything[7], means that $f^{-1}(f(x)) = x$ and $f(f^{-1}(y)) = y$. For example, the function

$$\sin(\pi x/2),$$

---

[6] Naturally, if f is defined on an interval I and g on an interval J, then the number $g(x)$ must belong to I for each number x in J.

[7] It must, of course, be the case that one cannot have $f(x) = f(x')$ for two distinct numbers x and x'. Here we have a simple interpretation of the graph of f^{-1}: it is obtained by *symmetry* with respect to the line $y = x$ (called "the first bisecting line").

defined in the interval $[0,1]$ and taking its values in this interval, has as its "inverse" the function

$$\frac{2}{\pi}\arcsin x\,,$$

defined on the same interval.

D. Permutations and Substitutions

The problems of "combinatorics", which appeared in relation to the calculation of probabilities from the sixteenth century on, and, in the eighteenth century, in the theory of determinants, led to the consideration of the different ways of arranging finitely many objects in order. These arrangements were called *the permutations* of the objects. For example, for three objects a, b, c, there are six permutations:

$$abc,\quad acb,\quad bac,\quad bca,\quad cab,\quad cba.$$

In general, for n distinct objects $x_1, x_2, ..., x_n$, it is easy to show by induction on n that the number of their permutations is the product

$$1 \cdot 2 \cdot 3 \cdot ... \cdot n,$$

which is written as $n!$ and is called *"factorial n"* (Appendix 2, §A).

When $x_1, x_2, ..., x_n$ are distinct numbers (real or complex), it is easy to construct an algebraic equation of degree n of which these numbers are the roots: that is,

$$P(t) \equiv (t - x_1)(t - x_2)...(t - x_n) = 0.$$

If we write

(19) $$P(t) = t^n - a_1 t^{n-1} + a_2 t^{n-2} - ... + (-1)^n a_n,$$

it is obvious that the coefficients $a_1, a_2, ..., a_n$ do not change when the factors $(t - x_1), (t - x_2), ...$ are arranged in a different order. These coefficients are *polynomials* in $x_1, x_2, ..., x_n$ which do not change when the sequence

$$x_1, x_2, ..., x_n$$

is replaced by *any permutation* of this sequence. This can easily be verified by a direct calculation, which gives

$$a_1 = x_1 + x_2 + ... + x_n$$
$$a_2 = x_1 x_2 + x_1 x_3 + ... + x_{n-1} x_n$$
$$\dotfill$$
$$a_n = x_1 x_2 ... x_n.$$

In general, a_k is the sum of all the products which are obtained by taking k of the numbers $x_1, ..., x_n$ in every possible way and multiplying them. This is because, in order to create the product $P(t)$, it is necessary to choose in every possible way one term in each of the factors $t - x_1, ..., t - x_n$, and add together all the products of these terms.

It has been known since the seventeenth century that each *symmetric* polynomial $S(x_1, x_2, ..., x_n)$ (that is, a polynomial which is unchanged by any permutation of $x_1, x_2, ..., x_n$) can be written in the form $R(a_1, a_2, ..., a_n)$, a polynomial in $a_1, a_2, ..., a_n$ whose coefficients are rational numbers if those of $S(x_1, ..., x_n)$ are rational. For example, for $n = 3$:

$$x_1^2 + x_2^2 + x_3^2 = (x_1 + x_2 + x_3)^2 - 2(x_1 x_2 + x_2 x_3 + x_3 x_1)$$
$$= a_1^2 - 2a_2\,,$$

$$x_1^2 x_2 + x_1^2 x_3 + x_2^2 x_1 + x_2^2 x_3 + x_3^2 x_1 + x_3^2 x_2$$
$$= (x_1 + x_2 + x_3)(x_1 x_2 + x_2 x_3 + x_3 x_1) - 3 x_1 x_2 x_3$$
$$= a_1 a_2 - 3 a_3\,.$$

In 1770 Waring gave an explicit formula for $R(a_1, a_2, ..., a_n)$ in the case where $S(x_1, x_2, ..., x_n)$ is known.

We have already recalled (see A, above) the sensational discovery of the resolution of the cubic equation by Scipione del Ferro. About 1540, L. Ferrari, a pupil of Cardan, obtained likewise the resolution of the fourth-degree equation, by a far more complicated formula in which, as in formula (7), superpositions of radicals were used. For two hundred years, however, all the endeavours of the best mathematicians to obtain an analogous formula to resolve the fifth-degree equation led only to failure.

In 1770 Lagrange decided to seek the reason for these repeated set-backs by linking the problem to questions of the *permutations* of the roots of an equation. All the methods of solving a fourth-degree equation were reached by way of a prior solution of an auxiliary cubic equation. It occurred to Lagrange to explain this fact by taking as his point of departure the following remark: if x_1, x_2, x_3, x_4 are the roots of the equation

$$t^4 - a_1 t^3 + a_2 t^2 - a_3 t + a_4 = 0\,,$$

then the polynomial

$$x_1 x_2 + x_3 x_4$$

is *not* symmetric, but when x_1, x_2, x_3, x_4 are permuted in $4! = 24$ different ways, we obtain not twenty-four different polynomials but *only three*, namely

$$s_1 = x_1 x_2 + x_3 x_4\,, \qquad s_2 = x_1 x_3 + x_2 x_4\,, \qquad s_3 = x_1 x_4 + x_2 x_3\,.$$

The elementary symmetric functions

$$s_1 + s_2 + s_3\,, \qquad s_1 s_2 + s_2 s_3 + s_3 s_1\,, \qquad s_1 s_2 s_3$$

are themselves symmetric polynomials of x_1, x_2, x_3, x_4, and therefore polynomials b_1, b_2, b_3 in a_1, a_2, a_3, a_4, and s_1, s_2, s_3 are the roots of the auxiliary cubic equation

$$t^3 - b_1 t^2 + b_2 t - b_3 = 0$$

which is expressed by Scipione del Ferro's formula. From here, at last, it is easy to deduce x_1, x_2, x_3, x_4 by means of the resolution of three quadratic equations (Appendix 1).

Once he had produced the solution of a fourth-degree equation, Lagrange obviously concluded from this that, if one could likewise find a polynomial $F(x_1, x_2, x_3, x_4, x_5)$ in relation to the five roots of a fifth-degree equation, which takes only *three or four values* when the five roots are permuted in $5! = 120$ ways, one could solve "by radicals" the fifth-degree equation. But he had doubts as to the existence of such a polynomial.

At the beginning of the nineteenth century, a pupil of Lagrange, the Italian P. Ruffini, effectively proved that such a polynomial *cannot exist*. By some very lengthy arguments which we cannot summarize here, he studied the systems S of all the permutations of x_1, x_2, x_3, x_4, x_5 for which a given polynomial $P(x_1, x_2, x_3, x_4, x_5)$ remains *invariant*. He showed that the number p of permutations of such a system divides 120, and that the number of distinct polynomials which can be formed from P by applying to it the 120 permutations of the x_j is $120/p$. Finally he showed that $120/p$ can never be equal to 3 or 4. This did not prove conclusively that it is impossible to solve an arbitrary fifth-degree equation by the superposition of radicals: Ruffini was hindered from reaching this conclusion for lack of ideas relating to what much later came to be called *fields* of numbers (see §3, C), thanks to which Abel, in 1824, was able to complete the proof.

What is important here for our purposes is the way in which, some years after Ruffini, the young Cauchy, in order to generalize these results, presented the concept of permutation from a completely new angle, which subsequent developments would show to be crucial. If, for example, we take four objects a, b, c, d arranged in that order, and if c, b, d, a is a permutation of these, Cauchy establishes for this permutation the *law*

$$a \mapsto c, \quad b \mapsto b, \quad c \mapsto d, \quad d \mapsto a$$

which associates with each object in the first place the corresponding object in the second place. He called this law a *substitution*, and he wrote it as

$$\begin{pmatrix} a & b & c & d \\ c & b & d & a \end{pmatrix}.$$

He understood, however, that the order in which the four objects are arranged is of no importance for the definition of a substitution, which could just as well be denoted

In particular,

$$\begin{pmatrix} a & b & c & d \\ c & b & d & a \end{pmatrix}$$

is the *same* substitution as

$$\begin{pmatrix} c & b & d & a \\ d & b & a & c \end{pmatrix}.$$

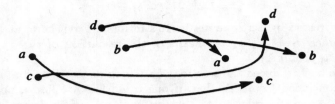

Fig. 37.

Cauchy did not limit himself to four or five objects, but set out to deal with the case of *any* number of objects. Although he suggested a general notation

$$\begin{pmatrix} a & b & c & \dots & l \\ \alpha & \beta & \gamma & \dots & \lambda \end{pmatrix}$$

for an arbitrary substitution, it is clear that a "picture" of this kind is practicable only for a small number of objects, and in working out his argument Cauchy used the abridged notation

$$\begin{pmatrix} A \\ B \end{pmatrix},$$

where A and B are any permutations of the objects in question.

Cauchy's most important innovation was to *compose* two substitutions: he wrote

$$\begin{pmatrix} A \\ B \end{pmatrix} \begin{pmatrix} C \\ D \end{pmatrix}$$

for the substitution obtained by first making the substitution $\begin{pmatrix} A \\ B \end{pmatrix}$, then the substitution $\begin{pmatrix} C \\ D \end{pmatrix}$. Since $\begin{pmatrix} C \\ D \end{pmatrix}$ can also be written as $\begin{pmatrix} B \\ E \end{pmatrix}$ for some permutation E, we have

$$\begin{pmatrix} A \\ B \end{pmatrix} \begin{pmatrix} B \\ E \end{pmatrix} = \begin{pmatrix} A \\ E \end{pmatrix} ;$$

for example,

$$\begin{pmatrix} a & b & c & d \\ b & a & c & d \end{pmatrix} \begin{pmatrix} b & d & c & a \\ d & a & b & c \end{pmatrix} = \begin{pmatrix} a & b & c & d \\ b & a & c & d \end{pmatrix} \begin{pmatrix} b & a & c & d \\ d & c & b & a \end{pmatrix}$$
$$= \begin{pmatrix} a & b & c & d \\ d & c & b & a \end{pmatrix}.$$

Cauchy also considered the composition of several substitutions,

$$\begin{pmatrix} A \\ B \end{pmatrix} \begin{pmatrix} C \\ D \end{pmatrix} \begin{pmatrix} E \\ F \end{pmatrix} \begin{pmatrix} G \\ H \end{pmatrix},$$

etc., and, in particular, he used the notation

$$\binom{A}{B}^k$$

for the substitution composed of k substitutions all equal to $\binom{A}{B}$. He was thus led to write

$$\binom{A}{A}$$

for the *"identity substitution"*, in which each object remains invariant. Although he does not say so explicitly, Cauchy cannot have failed to notice that

$$\binom{A}{B}\binom{B}{A} = \binom{A}{A} .$$

To see that there is always a natural number N so large that

$$\binom{A}{B}^N = \binom{A}{A} ,$$

we argue as follows. Since there is only a finite number of permutations, when all the substitutions

$$\binom{A}{B}^m$$

are formed in succession, *there must necessarily be two distinct natural numbers h and k such that*

$$\binom{A}{B}^h = \binom{A}{B}^k ,$$

and it can be supposed that $h > k$. From this we deduce that

$$\binom{A}{B}^h \binom{B}{A}^k = \binom{A}{B}^k \binom{B}{A}^k .$$

But we can write

$$\binom{A}{B}^p \binom{B}{A}^q = \binom{A}{B}^{p-1} \binom{A}{B}\binom{B}{A}\binom{B}{A}^{q-1} = \binom{A}{B}^{p-1} \binom{B}{A}^{q-1}$$

because

$$\binom{A}{B}\binom{B}{A}$$

is the identity. Hence we conclude that

$$\binom{A}{B}^{h-k} = \binom{A}{A} .$$

For us, the analogy with the composition of functions (§C) springs to mind at once. But for the mathematicians of the beginning of the nineteenth century, there were too many differences between a finite set and a straight line for a unification to be conceivable; this was to come only with Dedekind and Frege.

E. Displacements and Affinities

The idea of the movement of a rigid figure in the plane or in space is obviously something that everybody understands; but it is not certain that the Greeeks had had any notion of *displacement*: that is to say, the transition of a figure from its original position to its final one without taking account of its intermediate positions. The "equality" of two triangles in Euclid (in the sense of what Hilbert was to call their "congruence")[8] doubtless meant initially the possibility of carrying out a *movement* to put one on top of the other[9]. But to consider a displacement *in itself* as a mathematical object, independent of the "figures" which are displaced, is a way of thinking which scarcely made its appearance before the eighteenth century[10].

At that time, Euler explained the idea of a displacement as a "transformation" applied "en bloc" to the *whole space*, that is, a law which associates *every* point of the space (not only the points of a "figure") with another point of the space. Two such transformations can obviously be "composed" by performing first one and then the other. Euler shows that every displacement in space which leaves a point *fixed* is a *rotation* around a line passing through this point, and that *every* displacement is obtained by combining a rotation with a translation. In a plane marked off by two mutually perpendicular axes each displacement may be represented by associating with each point (x, y) the point with coordinates

$$(20) \qquad X = a + x \cos\theta - y \sin\theta, \qquad Y = b + x \sin\theta + y \cos\theta,$$

where a, b, and θ are real numbers characterizing the displacement. More generally, the name *affine* transformations is given to those transformations which are in a similar way defined by

$$(21) \qquad X = a + \alpha x + \beta y, \qquad Y = b + \gamma x + \delta y,$$

where the constants $a, b, \alpha, \beta, \gamma$, and δ are arbitrary real numbers. Euler already had in mind the particular cases

---

[8] Euclid uses the word "equal" in several senses: sometimes for two figures which can be superimposed, sometimes for two figures with the same area.

[9] Euclid, in the two cases of equality of triangles which he "proves" in this way (*Elements*, Book I, 4 and 8) says simply that one triangle "is applied" on the other (ἐφαρμόζεσθαι). This is one more example of his drawing on concepts which he has not defined and which he has not made the subject of any postulate. In the sixteenth century one of his commentators, J. Peletier, deplored this intrusion of methods which he thought brought discredit on "the dignity of geometry", and was already thinking that Euclid ought to have taken these propositions as postulates (this is what Hilbert was to do for one of them). Later Euclid avoided resorting to arguments of this kind, even where this would have been natural.

[10] We do however find in Descartes the idea of "an infinitely small rotation".

$$X = \alpha x, \qquad Y = \beta y,$$

and the geometers of the nineteenth century of made much use of *the homothety*, the case where $\alpha = \beta$, a concrete "picture" of which can be drawn by means of a *pantograph*, something used by draughtsmen to enlarge or reduce a figure keeping the proportions constant (Figure 38). The rulers pivot around O and are articulated at A, B, A', and B', forming two rhombuses with sides a and a'; the points M and M' are aligned with O and are such that $\overline{OM'}/\overline{OM} = -a'/a$, in such a way that when M describes a curve C, M' describes the homothetic curve, in the proportion $-a'/a$.

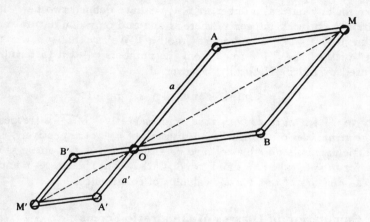

Fig. 38.

F. Calculation of Congruences of Natural Numbers

Many arithmetical questions (see for example Chapter IV, Appendix 3) have to do with integers (positive, negative or zero) divisible by a *given* natural number $m > 1$, that is to say, integers of the form km, where k is an integer. More generally, the integers $a + km$ are considered, where a and m are fixed and k is any integer (it could be said that this is an arithmetical progression "in both directions"). Since the time of Gauss, it has become customary, instead of saying that x is of the form $a + km$, to write

$$(22) \qquad x \equiv a \pmod{m},$$

which is read as "x is congruent to a *modulo* m": we say that a relation (22) is a *congruence* of numbers modulo m.

This notation obviously recalls the relation of equality $x = a$. This similarity of notation is entirely justified: in fact, from the two relations

$$(23) \qquad x \equiv a \pmod{m}, \qquad y \equiv b \pmod{m},$$

it can be deduced that

(24) $$x + y \equiv a + b \pmod{m}$$

and

(25) $$xy \equiv ab \pmod{m},$$

for it is only necessary to calculate, for $x = a + hm$ and $y = b + km$, that

$$x + y = a + b + (k + h)m, \qquad xy = ab + (hb + ka + hkm)m.$$

Hence for any natural number $m > 1$, we have defined two new "laws of composition". For each integer x, there is one and only one natural number r such that $0 \leq r \leq m - 1$ and $x \equiv r \pmod{m}$. We say that r is the "representative" of the "class" $\bar{r}$ formed from the numbers $r + km$ (a terminology already used by Euler). Naturally, however, if

$$0 \leq r \leq m - 1 \quad \text{and} \quad 0 \leq s \leq m - 1,$$

we can have $r + s > m$ or $rs > m$: the "representative" of $\overline{r + s}$ (respectively $\overline{rs}$) is the *remainder* of the Euclidian division of $r + s$ (respectively, rs) by m. In particular, it is possible to have $rs \equiv 0$ even if $r \not\equiv 0$ and $s \not\equiv 0$. For example, for $m = 6$, take $r = 2$ and $s = 3$. We say that the classes $\bar{r}$ and $\bar{s}$ are in this case *divisors of zero* for the calculus of congruences.

G. The Calculation of Classes of Quadratic Forms

The algebraic identity

(26) $$(a^2 + b^2)(c^2 + d^2) = (ac - bd)^2 + (ad + bc)^2$$

no doubt dates from classical times. Lagrange uses the more general identity

(27) $$(x^2 + Dy^2)(u^2 + Dv^2) = (xu - yvD)^2 + D(xv + yu)^2,$$

which actually originated with a Hindu mathematician, Brahmagupta (seventh century). Lagrange deduces from this that, if the equations

$$x^2 + Dy^2 = n_1$$
$$u^2 + Dv^2 = n_2$$

(where D is an integer) have integer solutions (x, y) and (u, v) (Chapter IV, §2, A), then the same is true of the equation

$$X^2 + DY^2 = n_1 n_2.$$

Gauss discovered a far more elusive law than the identity (27). It is not true in general that we have an identity of the form

$$(ax^2 + bxy + cy^2)(au^2 + buv + cv^2) = aX^2 + bXY + cY^2,$$

where X and Y are polynomials of the first degree in x, y and of the first degree in u, v:

$$(28) \quad \begin{cases} X = p'xu + q'yv + r'xv + s'yu \\ Y = p''xu + q''yv + r''xv + s''yu \end{cases}$$

with the coefficients p', q', ..., r'', s'' being integers. Nevertheless, Gauss was able to show, by means of a very laborious calculation, that, if $ax^2 + bxy + cy^2$ and $a'x^2 + b'xy + c'y^2$ are two binary[11] quadratic forms with integer coefficients which have the *same discriminant* Δ, then there is always a change of variables with coefficients $\alpha, \beta, \gamma, \delta$ such that $\alpha\delta - \beta\gamma = 1$,

$$\begin{cases} z = \alpha x + \beta\gamma \\ t = \gamma x + \delta y, \end{cases}$$

having the property that, if $az^2 + bzt + ct^2 = Ax^2 + Bxy + Cy^2$, then we have an identity

$$(Ax^2 + Bxy + Cy^2)(a'u^2 + b'uv + c'v^2) = A'X^2 + B'XY + C'Y^2,$$

where X, Y are given by formulae (28) with integer coefficients. Further, the form $A'X^2 + B'XY + C'Y^2$ also has the discriminant Δ; and all the forms obtained in this way are *equivalent*, their *class* (Chapter IV, §2, A) depending only on the *classes* C and C' of the forms $ax^2 + bxy + cy^2$ and $a'x^2 + b'xy + c'y^2$. Gauss then says that the class C'' is *composed* of the classes C and C', and he writes $C'' = C + C'$.

2. The First Structures

A. The Principal Properties of the Laws of Composition

The examples given in §1 are all concerned with what are known as the *laws of composition*: these can be summarily described as a procedure which makes it possible to effect a correspondence between two objects a and b belonging to the same set and a third well-determined object c of the set: this can be expressed in writing as $c = a \bowtie b$, where $\bowtie$ is a sign characterizing the law being considered. The most frequently used signs are $+$, $\cdot$, $\times$, $\circ$, $\star$, $\cup$, $\cap$, $\wedge$, and $\otimes$.

Euclid (*Elements*, Book VII, 16) had already thought that, for the product of two natural numbers a and b, the fact that $a \times b = b \times a$ needed to be proved. This is what came to be called, at the beginning of the nineteenth century, the *commutativity* of multiplication, and the same question at once arose for the new laws of composition which were then being formulated. It is immediately

---

[11] A homogeneous polynomial (or "form") is called "binary" if it contains only two variables, "ternary" if it contains three, etc.

evident that for Cauchy's "substitutions" (§1, D), there is no commutativity in general: in other words, if we use a single letter to denote a substitution (as does Galois), it is perfectly possible to have $TS \neq ST$: it happens, for example, if we take

$$S = \begin{pmatrix} 1 & 2 & 3 \\ 3 & 1 & 2 \end{pmatrix}, \qquad T = \begin{pmatrix} 1 & 2 & 3 \\ 2 & 1 & 3 \end{pmatrix},$$

for we have

$$ST = \begin{pmatrix} 1 & 2 & 3 \\ 3 & 2 & 1 \end{pmatrix}, \qquad TS = \begin{pmatrix} 1 & 2 & 3 \\ 1 & 3 & 2 \end{pmatrix}.$$

The same goes for a composition of two functions of a real variable: for example, if $f(x) = x^2$ and $g(x) = x + 1$, then we have

$$f(g(x)) = (x+1)^2, \qquad g(f(x)) = x^2 + 1.$$

The composition of displacements in the plane is not commutative either: a translation $T : (x, y) \mapsto (x + 1, y)$ and a rotation $R : (x, y) \mapsto (-y, x)$ do not commute.

On the other hand, in the calculus of congruences of numbers (§1, F), we have

$$a + b \equiv b + a \quad (\mathrm{mod}\ m)$$
$$ab \equiv ba \quad (\mathrm{mod}\ m).$$

In the same way, Gauss showed that $C' + C = C + C'$ for two classes of quadratic forms with the same discriminant.

Another property of a law of composition is the existence of what is called a *neutral element* : if we denote the law as $a \bowtie b$, then the neutral element is an element e such that

$$e \bowtie a = a \bowtie e = a$$

for every element a. If such an element exists, it is unique. The most obvious examples are 0 for the addition of real numbers, and 1 for their multiplication. For the functions of a real variable, the "identical" function $e(x) = x$ is the neutral element for the law $f \circ g$; and in the same way, for the substitutions (respectively, displacements in the plane) there is a neutral element, the substitution (respectively, displacement) "identity", which leaves all the objects (respectively, all the points) *fixed*. The existence of a neutral element for the composition of classes of quadratic forms with discriminant Δ is much less obvious: Gauss proved that, if $\Delta \equiv 0 \pmod 4$, then the neutral element E is the class of the form

$$x^2 - \frac{1}{4}\Delta y^2,$$

and that, if $\Delta \equiv 1 \pmod 4$, then it is the class of the form

$$x^2 + xy - \frac{1}{4}(\Delta - 1)y^2.$$

The existence of a neutral element is moreover not always guaranteed: for example, the product of two even integers is an even integer, but there is no neutral element for the multiplication of even integers.

Once a neutral element exists we can ask if there is an "inverse" a' of an element a: that is to say, such that

$$a \bowtie a' = a' \bowtie a = e \, .$$

It has long been known that a fraction m/n has an "inverse" n/m for multiplication. The introduction of negative numbers made it possible to say that each positive integer n has an "inverse" $-n$ for addition, but here the term "*negative*" is used. We have already remarked that a substitution $\begin{pmatrix} A \\ B \end{pmatrix}$, in Cauchy's notation, has an "inverse" substitution $\begin{pmatrix} B \\ A \end{pmatrix}$ for composition; and in the same way every displacement in the plane has an "inverse", the displacement which brings every point back to its original position. On the other hand, for the real functions of a variable defined on $\mathbf{R}$, the function $f(x) = x^3$ has an "inverse" function $g(x) = x^{1/3}$ for the composition of functions, but the same is not valid for $f(x) = x^2$, for there can be no real function g such that $g(x)^2 = x$ when x is negative. For the congruences of numbers (§1, F), there is not necessarily any inverse for their multiplication: for example, there is no natural number n such that $2n \equiv 1 \pmod 4$, and therefore 2 has no inverse "modulo 4" (Appendix 3, §A). On the other hand, Gauss was able to show in addition that every class C of quadratic forms has an "inverse" C' such that $C + C' = E$. (We still prefer to say the "class *opposite* to C" for C'; the term "opposite" generally replaces "inverse" when the law of composition is denoted by $+$.)

There is a property which was not explicitly pointed out until the beginning of the nineteenth century, the *associativity* of a law of composition, which, with the general notation $\bowtie$, is written

$$(a \bowtie b) \bowtie c = a \bowtie (b \bowtie c).$$

No doubt it was because this property held for all the known laws that it had not been mentioned. For the addition and the multiplication of real numbers, it was used automatically, and for the composition of functions (or of substitutions, or of displacements), it is just as "obvious", since $(f \circ g) \circ h$ and $f \circ (g \circ h)$ are functions whose value at x is in both cases $f(g(h(x)))$. The composition of classes of quadratic forms is the only case which requires a proof, and this is what Gauss supplied explicitly. It was not until 1845 that *non-associative* laws were defined.

Finally, when there are *two* laws of composition on the same objects, the property of *distributivity* makes its appearance; for example

$$a(b + c) = ab + ac$$

for the multiplication of numbers with respect to their addition. This was also noted explicitly at the beginning of the nineteenth century. It is not, however, always satisfied: for example, for the composition of real functions with respect to their addition we do indeed have

$$(f+g) \circ h = (f \circ h) + (g \circ h),$$

but on the other hand

$$f \circ (g+h) \neq (f \circ g) + (f \circ h)$$

in general.

B. Groups of Transformations

We have seen (§1, D) that Cauchy, in 1815, introduced the composition ST of two "substitutions" defined for any *finite* set of objects. As is indicated by the title of his monograph[12], the question at the centre of his concerns is that which had already been broached by Lagrange and Ruffini: the determination of the substitutions on n variables $x_1, x_2, ..., x_n$ which leave *invariant* a given polynomial $P(x_1, x_2, ..., x_n)$. It follows immediately (no matter what polynomial P we have) that, if S and T are two substitutions with this property, then the same is true for ST and for S^{-1}. Furthermore, if U is a substitution which transforms P into a different polynomial Q[13], then, by definition, if S leaves P invariant, SU also transforms P into Q. Conversely, if the substitution V also transforms P into Q, then $S = VU^{-1}$ leaves P invariant, and hence V can be written as $V = V(U^{-1}U) = (VU^{-1})U = SU$. Cauchy considered *all* the different polynomials $P = P_1, P_2, ..., P_d$ obtained by *all* the $n!$ possible substitutions: let $U_1 = I, U_2, ..., U_d$ be substitutions which transform P respectively into $P_1 = P, P_2, ..., P_d$, and on the other hand let $S_1 = I, S_2, ..., S_m$ be all the substitutions which leave P invariant. Then he divides all the $n!$ substitutions into *classes* in the following way:

$$
\begin{aligned}
&U_1 = I = S_1, \quad S_2, S_3, ..., S_m \qquad && P \text{ left invariant} \\
&U_2, \ S_2U_2, \ ..., S_mU_2 \qquad && P \text{ transformed into} P_2 \\
&\qquad\qquad \cdots\cdots\cdots\cdots\cdots\cdots \\
&U_d, \ S_2U_d, \ ..., S_mU_d \qquad && P \text{ transformed into} P_d.
\end{aligned}
$$

Since each row has the same number m of terms, we have

(29) $$n! = md,$$

---

[12] *Dissertation on the Number of Values a Function can Acquire when the Quantities it Includes are Permuted in All Possible Ways.*

[13] For example, the substitution

$$\begin{pmatrix} x_1 & x_2 & x_3 & x_4 \\ x_2 & x_4 & x_1 & x_3 \end{pmatrix}$$

transforms the polynomial $x_1x_2 + x_3x_4$ into $x_2x_4 + x_1x_3$.

a result in fact already obtained by Lagrange, except for the terminology. It is from this result that Cauchy proceeds, by means of ingenious combinations of substitutions, to generalize Ruffini's result by showing that, if n is a prime, we necessarily have $d = 2$ or $d \geq n$.

One further step in abstraction was taken by Galois in 1830, when, instead of continuing to talk of invariant polynomials, he uses the term *group of substitutions*[14] on n objects for a *subset* Γ of the $n!$ substitutions such that, if S and T are in Γ, the same is true of their "product" ST. Since this means, in particular, that S^k is in Γ for each exponent k , we can write $SS^{k-1} = S^{k-1}S = I$, where $S^k = I$ It follows that S^{k-1} is the *inverse S^{-1}* of S, which is therefore also in Γ. The totality $\mathfrak{S}_n$ of the $n!$ substitutions is obviously a group, called the *symmetric group* on n objects. If Γ is a group of substitutions and Γ' another group *contained* in Γ, we say that Γ' is a *subgroup* of Γ. The *order* of a group of substitutions is by definition the number of its elements: if Γ' is a subgroup of Γ, and g and g' are the orders of Γ and Γ', respectively, then g' is a *divisor* of g.

If S is any substitution, then there exists a *smallest* number h such that $S^h = I$. Hence the substitutions $I, S, S^2, ..., S^{h-1}$ are all distinct, for if we had $S^p = S^q$ for $1 \leq p < q < h$ we would deduce that $(S^{-1})^p S^p = (S^{-1})^p S^q$, that is to say, $I = S^{q-p}$ with $q - p < h$, contradicting the definition of h. These h substitutions form a subgroup because $S^p S^q = S^{p+q}$ and, if $h \leq p + q < 2h$, we have $S^{p+q} = S^h S^{p+q-h}$. We say that such a subgroup is *cyclic*, and that it is *generated* by S, and we also say that h is the *order* of S.

An example of a cyclic group is formed by the powers of what is known as a *circular* substitution

$$\begin{pmatrix} 1 & 2 & 3 & ... & n-1 & n \\ 2 & 3 & 4 & ... & n & 1 \end{pmatrix}.$$

If the numbers $1, 2, ..., n$ at the vertices of a regular polygon of n sides be placed on a circle, S can be represented as a rotation in which each vertex moves onto the next one, and the last one, n, returns to the first vertex (Figure 39). It follows at once that, for $1 \leq p < n$, we have

$$S^p = \begin{pmatrix} 1 & 2 & 3 & ... & n-1 & n \\ p+1 & p+2 & p+3 & ... & p-1 & p \end{pmatrix},$$

and hence the substitutions S^p for $1 \leq p < n-1$ are all distinct, and $S^n = I$. This can be written briefly as $S = (1\ 2\ 3\ ...\ n)$.

Examples of Subgroups

The six elements of the symmetric subgroup $\mathfrak{S}_3$ are written, using the previous notation, as

---

[14] Galois sometimes uses the term "permutation" in the sense in which Cauchy uses "substitution". Later it was the term "group of permutations" which was to prevail.

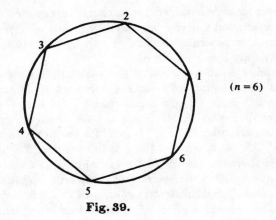

Fig. 39.

$$I, \quad (1\ 2), \quad (1\ 3), \quad (2\ 3), \quad (1\ 2\ 3), \quad (1\ 2\ 3)^2,$$

where

$$(1\ 2\ 3)^2 = \begin{pmatrix} 1 & 2 & 3 \\ 3 & 1 & 2 \end{pmatrix}.$$

There is a cyclic subgroup $\mathfrak{A}_3$ of order three, formed by

$$I, \quad (1\ 2\ 3), \quad (1\ 2\ 3)^2.$$

Then there are three subgroups of order two

$$\{I, (1\ 2)\}, \quad \{I, (1\ 3)\}, \quad \{I, (2\ 3)\}$$

And finally there is the subgroup consisting only of the neutral element I.

It was these concepts, used with much ingenuity, which enabled Galois to solve completely the problem of the resolution of an algebraic equation "by radicals" by showing that it is equivalent to a problem about what is now called the "Galois group" of the equation (Appendix 2, §B). It was also Galois theory and the development of group theory which made it possible to determine all the geometrical problems which can be solved "with ruler and compass". What must particularly be emphasized is that it was *necessary*, in order to reach the solution to these problems, resolutely to *leave behind* classical algebra and to introduce new "abstract" objects, for that is what these groups of substitutions are, of a completely different nature from algebraic equations. This is the same procedure as was to be followed later by Dirichlet and Riemann when they broached the study of prime numbers by introducing analytic functions (Chapter IV, §2, B). Here must be seen the earliest manifestations of the profound unity of mathematics, breaking through the traditional partitions.

Some years after Galois' death, crystallographers became interested in a geometrical problem that seemed far removed from the theory of equations: the determination of the rotations in space which leave a given polyhedron *invariant*. For example, in the case of a cube, these rotations must all leave the centre of the cube fixed, and so their axes all pass through this point. Further, the rotations permute the eight vertices among themselves, and the product of two of them is again a rotation which leaves the cube invariant. Thus they form a *subgroup* Γ of the group $\mathfrak{S}_8$ of permutations of the eight vertices. In fact, it is a good exercise to ascertain that Γ has order 24 (whereas $\mathfrak{S}_8$ has order $8! = 40\,320$) and to determine explicitly the product of two rotations of Γ (Appendix 2, §E). But the kinship of these groups with Cauchy's and Galois' groups of substitutions does not appear to have been noticed at first, and it was only belatedly that crystallographers made use of the language of groups.

In another area, the geometry of a plane or of space, which made great advances in the first half of the nineteenth century, the "changes of variables" of Lagrange and Gauss (Chapter IV, §2, A) came to be interpreted as "transformations". These are particular cases of the changes which, for example, effect a correspondence between a point in the plane of coordinates (x, y) and the point of coordinates (X, Y) given by

$$(30) \qquad S \; : \; \begin{cases} X = a + \alpha x + \beta y \\ Y = b + \gamma x + \delta y, \end{cases}$$

where the coefficients $a, b, \alpha, \beta, \gamma, \delta$ are arbitrary real numbers. These are what are called "affine transformations" of the plane (§1, E). After Gauss and Eisenstein, they were denoted by a single letter. The "composition" (or product) ST of two of them is a third one, and if $\alpha\delta - \beta\gamma \neq 0$, (30) can be solved for x and y, giving an affine transformation S^{-1} which is the "inverse" of S. However it was scarcely before 1860 that these properties came to be expressed by saying that the transformations (30) with $\alpha\delta - \beta\gamma \neq 0$ form a group, the *affine group* $A(2)$. The displacements which are defined by formula (20) form a subgroup of this group. Another important subgroup is formed by the transformations (30) for which $a = b = 0$ and $\alpha\delta - \beta\gamma \neq 0$: this is the *general linear group of the plane* $GL(2, \mathbf{R})$, the members of which are denoted in an abridged form (already used by Gauss) by a table

$$\begin{pmatrix} \alpha & \beta \\ \gamma & \delta \end{pmatrix},$$

later called a *matrix*. The relations (30) are written as a single equation between the vectors $\begin{pmatrix} x \\ y \end{pmatrix}$ and $\begin{pmatrix} X \\ Y \end{pmatrix}$ (§1, B):

$$(31) \qquad \begin{pmatrix} X \\ Y \end{pmatrix} = \begin{pmatrix} \alpha & \beta \\ \gamma & \delta \end{pmatrix} \begin{pmatrix} x \\ y \end{pmatrix}$$

in such a way that, if

$$S = \begin{pmatrix} \alpha & \beta \\ \gamma & \delta \end{pmatrix} \quad \text{and} \quad T = \begin{pmatrix} \alpha' & \beta' \\ \gamma' & \delta' \end{pmatrix},$$

then we have[15]

(32)
$$ST = \begin{pmatrix} \alpha\alpha' + \beta\gamma' & \alpha\beta' + \beta\delta' \\ \gamma\alpha' + \delta\gamma' & \gamma\beta' + \delta\delta' \end{pmatrix}.$$

It must be observed that, in order for a subset Γ of $A(2)$ to be a subgroup, it is *not enough* for it to contain the identity

$$I = \begin{pmatrix} 1 & 0 \\ 0 & 1 \end{pmatrix}$$

and for the product of two transformations S and T of Γ to be also in Γ. For example, if

$$S = \begin{pmatrix} 1 & 1 \\ 0 & 1 \end{pmatrix},$$

then we have

$$S^k = \begin{pmatrix} 1 & k \\ 0 & 1 \end{pmatrix}$$

for each natural number k, and every product $S^h S^k = S^{h+k}$ is in Γ, but these transformations and the identity I do not form a group because the inverse

$$S^{-1} = \begin{pmatrix} 1 & -1 \\ 0 & 1 \end{pmatrix}$$

of S is not included in the set. (The set only forms what is known as a "semigroup").

C. "Abstract" Groups

We have seen (§1, G) that Gauss, in studying the law of composition $C_1 + C_2$ of classes of quadratic binary forms with integer coefficients and given discriminants, had proved that this law is commutative and associative, that it possesses a neutral element E, and that each class C has an "opposite" C' such that $C + C' = E$. Nowadays obviously we say that these classes form a *commutative group*. It is the same with the " congruence classes modulo m" (§1, F) with respect to the additive law

$$\overline{r} + \overline{s} = \overline{r + s}.$$

Here $\overline{0}$ is the neutral element and $\overline{-r}$ is the class "opposite" to $\overline{r}$. The m elements of the group are

$$\overline{1},\ \overline{2} = 2 \cdot \overline{1},\ \overline{3} = 3 \cdot \overline{1},\ \ldots,\ \overline{m-1} = (m-1) \cdot \overline{1},\ \overline{m} = \overline{0} = m \cdot \overline{1}.$$

This group is *cyclic* and *generated* by $\overline{1}$. For a long time, however, well after the concept of a group of *transformations* had become current, nobody seems

---

[15] Here ST means that the tranformation T is made first, and then the transformation S, in accordance with the notation $f \circ g$ for the composition of two functions (§1, C).

to have noticed the analogy. What is more, it seems that nobody before 1850 ever wrote that the real or complex numbers form a group for addition, or that the positive rational numbers form a group for multiplication! Here we put our finger on the difficulty which nineteenth-century mathematicians had in freeing themselves from the traditional conception of the division of mathematics into parts characterized by the mathematical *objects* studied: natural numbers for arithmetic, equations for algebra, space and figures for geometry, functions for analysis. It took the whole of the nineteenth century to break this rigid compartmentalization and to reach the modern conception in which it is the *relations* between objects which count.

In the case of groups, the theory of quadratic binary forms with integer coefficients belonged to arithmetic, and for a long time mathematicians working in geometry knew nothing about the work of contemporary arithmeticians. Even Kronecker and Cayley, who had an encyclopedic knowledge of mathematics, did not manage to work out completely the general definition of a group. It was not given until 1882 for finite groups or (more generally) those with a finite number of generators[16], and not until 1893 for completely general groups (see §3, C, I).

D. Quaternions and Algebras

We have seen that the concept of group was introduced as a thing *in itself* in the course of studying problems of very diverse origins, in which it was revealed as an underlying principle in a *natural* way. In the same way, the introduction of complex numbers was *unavoidable* once the wish arose to deepen the solution of algebraic equations. It can be said that these two theories are *motivated*.

It is not the same with the theory of *quaternions*, discovered by Hamilton in 1843. This is the first example in history and the prototype of a theory that introduces new objects which, at the time when they are defined, do not answer any need, but are brought into being solely out of curiosity, just "to see".

Calculations with complex numbers (§1, A) had been assimilated into a system of calculation with points of the *plane*. As soon as this idea appeared, several mathematicians were to be seen asking themselves the question: is there an analogous system of calculation with points in *space*? Until 1840 no serious attempt had been made to answer this question[17]. At that point, Hamilton, in pursuing reflections on calculations with complex numbers considered as pairs $\begin{pmatrix} a \\ b \end{pmatrix}$ of real numbers, attacked the problem in a systematic way for triples

---

[16] The elements $a_1, a_2, ..., a_p$ constitute a "system of generators" of a group G if every element of G is the product (possibly in many ways) of a sequence of elements $x_1 x_2 ... x_k$ (k being an arbitrary natural number) in which the x_j are equal to one of the a_i or to its inverse a_i^{-1}. We say also that the a_i *generate* G.

[17] An enigmatic phrase of Gauss's could mean that he did not believe such an extension to be possible.

$$\begin{pmatrix} a \\ b \\ c \end{pmatrix} .$$

Since, for complex numbers, the sum is defined by the addition of coordinates, Hamilton naturally took the definition of addition to be the same as that for vectors:

$$\begin{pmatrix} a \\ b \\ c \end{pmatrix} + \begin{pmatrix} a' \\ b' \\ c' \end{pmatrix} = \begin{pmatrix} a + a' \\ b + b' \\ c + c' \end{pmatrix} .$$

But the question was how to define the multiplication

$$\begin{pmatrix} a \\ b \\ c \end{pmatrix} \times \begin{pmatrix} a' \\ b' \\ c' \end{pmatrix} .$$

Hamilton naturally wanted the properties valid for the multiplication of real or complex numbers[18] still to be satisfied, that is: associativity, commutativity and distributivity with respect to addition. Taking as his model the complex numbers, he wrote

$$\begin{pmatrix} a \\ b \\ c \end{pmatrix} = a \cdot 1 + b \cdot i + c \cdot j$$

on setting

$$1 = \begin{pmatrix} 1 \\ 0 \\ 0 \end{pmatrix} , \quad i = \begin{pmatrix} 0 \\ 1 \\ 0 \end{pmatrix} , \quad j = \begin{pmatrix} 0 \\ 0 \\ 1 \end{pmatrix} ,$$

which he envisaged as three vectors of unit length lying on mutually perpendicular axes in space, just as 1 and i are two vectors of unit length on two mutually perpendicular axes in the plane used in the geometric representation of complex numbers.

The essential observation which guided all his endeavours was that a complex number $z = a + bi$ has associated with it the *length* $\sqrt{a^2 + b^2}$ of the vector $\begin{pmatrix} a \\ b \end{pmatrix}$. This length is denoted by $|z|$ (and is also called the *modulus* of z). We have, for the multiplication of complex numbers, the "law of moduli"

(34) $$|zz'|^2 = |z|^2 |z'|^2 ,$$

which in fact is nothing more than a particular case of Brahmagupta's identity (27)

(35) $$(a^2 + b^2)(x^2 + y^2) = (ax - by)^2 + (ay + bx)^2 .$$

---

[18] The fact that properties of addition and multiplication, where the order structure of **R** does not intervene, are the same for both **R** and its "extension" **C** was considered by the contemporary English mathematicians as a necessity, which they termed the "principle of permanence".

Hamilton's idea of an analogy with the calculation of complex numbers led him to associate the length $\sqrt{a^2 + b^2 + c^2}$ with his "triples" $a + bi + cj$, and he tried to see if he could find a multiplication

$$(a + bi + cj)(x + yi + zj) = A + Bi + Cj$$

which satisfies the same "law of moduli"

(36) $$(a^2 + b^2 + c^2)(x^2 + y^2 + z^2) = A^2 + B^2 + C^2$$

over and above the algebraic properties recalled above.

For $y = z = 0$, it is sufficient to take $A = ax$, $B = bx$, and $C = cx$ to satisfy (36). Hamilton therefore considered that it was justifiable to take

$$\begin{pmatrix} 1 \\ 0 \\ 0 \end{pmatrix}$$

as the neutral member in his multiplication.

If, in the second place, we take $c = z = 0$, relation (36) is satisfied by taking $A = ax - by$, $B = ay + bx$, and $C = 0$ according to (35). Hamilton therefore admitted that it was necessary to have $i^2 = -1$ for his multiplication; and in the same way by taking $b = y = 0$, we see that it is necessary to have $j^2 = -1$.

Admitting all the conditions that he imposed, there results:

(37) $\quad (a + bi + cj)(x + yi + zj) =$
$$= (ax - by - cz) + (ay + bz)i + (az + cx)j + (bz + cy)ij .$$

But what was to be done with ij, which should also have been a triple $\alpha + \beta i + \gamma j$? It is very instructive to follow at this point the tentative steps of a creative mathematician, recounted in detail as they are by Hamilton; they are much more accessible than those described by Poincaré [9] in his discovery of "Fuchsian" functions. Hamilton experiments successively with taking $ij = 1$, $ij = -1$, and even, in desperation, $ij = 0$. But in none of these cases do the values of A, B, C deduced from (37) verify (36), for we have

(38) $\quad (a^2 + b^2 + c^2)(x^2 + y^2 + z^2) =$
$$= (ax - by - cz)^2 + (ay + bx)^2 + (az + cx)^2 + (bz - cy)^2 .$$

However, in comparing this formula to (37), he noticed that the coefficient of the unknown element $k = ij$ would be equal to the last of the polynomials, $bz - cy$, if, in the development (by distributivity) of the left-hand side of (37), *he had taken* $ji = -ij$. Then he had the sudden intuition that the problem was capable of a solution, provided that:

1) we multiply *quadruples*

$$a + bi + cj + dk$$

and *not* triples, with the supplementary rule that $k^2 = -1$;

2) the commutativity of the multiplication be *abandoned*, whilst the associativity be maintained, as is the distributivity (on the left and the right) with respect to addition, taking

$$ij = -ji = k\,.$$

By associativity he deduced the further rules:

$$ik = i(ij) = (i^2)j = -j\,;$$
$$ki = (ij)i = -(ji)i = -j(i^2) = j\,;$$
$$jk = j(ij) = -j(ji) = -(j^2)i = i\,;$$
$$kj = (ij)j = i(j^2) = -i\,.$$

With these rules, he obtained in effect a "law of moduli" for what he called *quaternions* $a + bi + cj + dk$. It can be verified that

$$(39) \quad (a + bi + cj + dk)(x + yi + zj + tk) =$$
$$= (ax - by - cz - dt) + (ay + bx + ct - dz)i +$$
$$+ (az - bt + cx + dy)j + (at + bz - cy + dx)k\,,$$

and we have the identity

$$(40) \quad (a^2 + b^2 + c^2 + d^2)(x^2 + y^2 + z^2 + t^2) =$$
$$= (ax - by - cz - dt)^2 + (ay + bx + ct - dz)^2 +$$
$$+ (az - bt + cx + dy)^2 + (at + bz - cy + dx)^2,$$

which moreover had already been found by Euler[19].

Having thus brilliantly solved his problem, Hamilton set himself the question of what his quaternions could be "used" for, and the last twenty years of his life were devoted to this quest, with the help of some slightly over-enthusiastic disciples. It has to be admitted that their results remained somewhat scanty, and they did not succeed in communicating to the mathematical community of their day their conviction that the theory of quaternions was a panacea for all problems. It was only much later, at the end of the nineteenth century, that quaternions came up *naturally* in theories such as the linear representation of groups or the structure of Lie groups, theories which did not exist in Hamilton's day.

Another consequence of Hamilton's labours was the search for *all* possible "multiplications" between "triples" or "quadruples" and, more generally,

---

[19] Hamilton did not know of this work of Euler's. He was not acquainted either with an observation of Legendre's showing that an identity (36), in which A, B, C are linear functions of a, b, c and of x, y, z with rational coefficients, *cannot exist*: in fact, if we take $a = b = c = 1$, $x = 4$, $y = 2$, and $z = 1$, then the number $63 = 3 \times 21$ ought to be the sum of the squares of three rational numbers. But since 63 is of the form $8n + 7$, it has been known since Diophantus' time that this is impossible.

"vectors of n dimensions" (§1, B) which are possible if we no longer impose commutativity or even associativity, but keep distributivity (on the right and left) with respect to the addition of vectors. The chief merit of this research has been to show the variety of "calculations" which can be made on new "abstract" mathematical objects, in contrast with the classical attitude which tended to believe that the usual rules for algebraic calculation were intangible. But the investigation of the relations between these rather "gratuitous" algebraic researches and other areas of mathematics, or their applications to natural science, hardly began until after 1890. We shall return to this in §5, A, VI.

3. The Language of Sets and General Structures

A. The Concept of Set

The idea of bringing together objects of the same nature in a "collection" is no doubt as old as language, and the same is true for the concept of a "part" of a "whole", which can be a collection of objects or a "spread", such as a plane or a space. The need to understand and use these concepts became evident at the very beginning of Greek mathematics, in Euclid's famous "Common Notion", stating that "the whole is greater than the part", to which he appeals incessantly in his *Elements*, for comparing two lengths, two angles or two areas.

Mathematicians of all epochs moreover have not been able to avoid speaking of the sets formed by the objects they deal with and of the subsets of these sets, under various names: "lieux géométriques" ("geometric loci") of points in the plane satisfying a certain condition, "classes" of numbers or of quadratic forms in Euler and Gauss, "subgroups" for Cauchy and Galois, "Mannigfaltigkeiten" ("multiplicities") of any dimension in Riemann, "sets" of symbols in Cayley. German authors also used "Gesamtheit", "Inbegriff", "System", "Gebiet": the word "Menge", already used by Bolzano, only became predominant with Cantor.

Nevertheless, a modern reader may be surprised to find that the Greek mathematicians never spoke of the "set of natural numbers", nor of the "set of points of the plane", even though these concepts seem familiar to us. The reason is that these sets are "infinite", which is a purely negative notion, difficult to conceptualize, and it has been the subject of countless debates within the schools of philosophy (cf. Chapter VI, §3). No doubt this fact must be seen as one of the reasons for the unwillingness of mathematicians from Euclid to Cauchy to speak of this concept, for fear of becoming embroiled in controversies which they feared would be futile. When they are faced with an infinite set of objects, they either do as Euclid did with respect to prime numbers, rest content with saying that after a finite number have been enumerated there still remain some more; or else, as Gauss did for classes of binary quadratic forms, they say nothing at all, trusting to the intelligence of their readers. Bolzano seems to have been the first mathematician to speak freely of infinite sets; but it was not until Dedekind arrived on the scene that a real mathematical definition of these objects was given. This will be dealt with in detail in Chapter VI, §3.

B. The Language of Sets

It was moreover Dedekind[20] who, precisely with the aim of defining infinite sets and natural numbers, in his work *Was sind und was sollen die Zahlen*, which appeared in 1888 but had already been drafted in 1878, begins with a sort of "summary of results", in which he introduces a very precise language, in strong contrast with the vague usage of his contemporaries. Although the work did not have the immediate influence which it deserved, the need to adopt a similar and *uniform* language in all areas of mathematics became felt little by little at the beginning of the twentieth century. With certain additions subsequent to Dedekind, it has become what can be called the *naïve language of sets*, universally used in our day, and something which mathematicians could not do without except at the price of interminable circumlocutions.

Dedekind did not feel the need to present this language in axiomatic form: visibly, for him as for his contemporaries, the elementary definitions and results which he enumerates express "common-sense truths". It can be said that, in spite of the subsequent controversies of which we shall speak in Chapter VI, mathematicians nowadays still use this language in the same way, and what is important for them is that it enables them to express their ideas without ambiguity.

Dedekind does not, any more than Euclid does with respect to the point and the line, give a true definition of a set, nor of a member of this set: the fundamental relationship is that whereby an element x *belongs to* a set E, which we now denote as $x \in E$ (its negation being denoted as $x \notin E$), and which states that "x belongs to E" or "x is a member of E". The first concept to be defined is that of a *subset* (or *part*): A is a subset of E, and we now write $A \subset E$ (and, for its negation, $A \not\subset E$) if every member of A is also a member of E. Dedekind grants that, if we have at the same time $A \subset B$ and $B \subset A$, then $A = B$: a set is entirely determined by its members. He passes rapidly over the operations of union and intersection[21], which had long been known in particular cases and which had been codified by Boole in connection with formal logic[22] (Chapter VI, §5, A). He simply emphasizes that these operations can be applied to any *family* of subsets of a set (he

---

[20] It is only fair also to mention here the contributions of Frege, which date from the same period, and aim in the same way to introduce a very precise language "of sets".

[21] Unlike Boole, Dedekind does not speak of the *complement* of a subset A of a set E, namely the set of $x \in E$ which do not belong to A, often denoted as $E \setminus A$. Neither does he speak of the *empty* set, which it is useful to introduce for purposes of argument (Leibniz had already thought of this). It is now denoted $\emptyset$, and one can write $E \setminus E = \emptyset$.

[22] Dedekind does not refer to Boole, although he must have known his work.

has indicated clearly that the subsets of a set E are members of another set, which is now designated by $\mathfrak{P}(E)$).

What was new and what was to be essential for the whole of mathematics was the entirely general conception of *function*, or, as we prefer to say nowadays, of *mapping*. Instead of confining himself, as in previous conceptions, to real (or complex) functions of one or several real variables, Dedekind went all at once to the extreme limit of generalization: given *any* two sets, E and F, a mapping f of E into F is a *law* ("Gesetz") which provides for every element x in E a corresponding *well-determined*[23] element in F, that is, its *value* at x, which is denoted in a general way by $f(x)$ (other notations being used for particular mappings). We have now adopted the custom of writing $x \mapsto f(x)$ to denote a mapping f, which often avoids the necessity of introducing a new letter, as for example in writing $x \mapsto x^2$ for $x \in \mathbf{R}$. Use is also often made in current mathematics of the notations $f : E \to F$ or $E \xrightarrow{f} F$ to denote precisely the set E on which the function f is defined and the set F in which it takes its "values"[24].

Dedekind's exposition still lacks a notion in constant use, introduced only by Cantor a little later, that of the *product* (or "Cartesian product") $E \times F$ of any two sets. This is the set of *pairs* (x, y) for all members x of E and all members y of F[25], a natural and indispensable generalization of Cartesian coordinates. We can connect the concept of a mapping $f : E \to F$ with that of a subset of the product $E \times F$: the *graph* Γ_f of f is the subset of $E \times F$ formed by the pairs $(x, f(x))$ for all members x of E, an obvious generalization of the classical "graph" of a real function of a real variable (Chapter III, §8).

The product $E_1 \times E_2 \times \ldots \times E_n$ of any natural number n of sets is defined in the same way: the product of n sets all identical to E is written as E^n.

The importance of this language is that it enabled mathematicians from the latter years of the nineteenth century onwards to speak of relations between objects whose nature is *completely indeterminate*: they are simply members of sets posited as the basic objects of an axiomatic theory. A *structure* will be determined by a certain number of such "basic" relations, subject to a system of axioms; the theory of such a structure will be the progressive revelation of properties which are purely the consequence of its axioms, and do not depend on the nature of the mathematical objects which could be shown to satisfy these axioms. These properties will in the end constitute theories comparable in scope to the classical theories. In the next few pages we shall give some examples of these.

---

[23] Following Riemann, Dedekind thus eliminates the confused notions of "multivalued functions" (that is, functions which could take several values for the same value of the variable) which had been introduced into the theory of analytic functions of a complex variable in relation to the definition of the square root or the logarithm of a complex number.

[24] Hence this concept embraces the most general functions with real values, defined by Dirichlet, Cauchy's "substitutions", and all the geometric "transformations".

[25] Of course, if $E = F$, then we must distinguish (y, x) from (x, y) in the case where $x \neq y$.

C. Algebraic Structures

We shall begin with some examples of what are known as *algebraic structures*, chosen from among those which mathematicians nowadays use most, and which are also used in many applications to natural science and problems in technology.

C I) Groups

The oldest and simplest structure is that of a group. Examples have been given above (§2, B and C) of some groups which mathematicians were *compelled* to study before the general concept had been formulated.

Proceeding as Hilbert does in the case of geometry (Chapter III, §4), a *group* is formed from three "basic" objects: a set G, a member $e \in G$, and a *law of composition* in G, which is a *mapping* $m : G \times G \to G$. There are three axioms:

1) The law m is *associative*, that is to say,

$$(41) \qquad m(m(x, y), z) = m(x, m(y, z)) \quad \text{for} \quad x, y, z \in G.$$

2) The element e is the *neutral element* for m, that is to say,

$$(42) \qquad m(e, x) = m(x, e) = x \quad \text{for} \quad x \in G.$$

3) Each member $x \in G$ has an *inverse* x', or in other words we have

$$(43) \qquad m(x, x') = m(x', x) = e.$$

A group is said to be *commutative* if, in addition, we have

$$(44) \qquad m(x, y) = m(y, x) \quad \text{for} \quad x, y \in G.$$

We say that G is a *finite* group if G has a finite number of members. This number is called the *order* of the group.

A subset H of a group G is called a *subgroup* of G if, for x and y in H, we have $m(x, y) \in H$ and $x' \in H$ (product and inverse are in H). Obviously all this generalizes what we have seen for groups of substitutions (§2, B). The same arguments prove that, if G is finite and if H is a subgroup of G, then the order of H *divides* the order of G.

Usually we write xy instead of $m(x, y)$ and x^{-1} instead of x'. We define successively $x^2 = xx, x^3 = x^2 x, \ldots$; if there is a smallest natural number k such that $x^k = e$, then the elements $e, x, x^2, \ldots, x^{k-1}$ are all distinct and form a commutative subgroup of order k, said to be *cyclic*. We also say that k is the *order* of x. If G is finite, then all the elements of G are of finite order, and their orders divide the order of G. Thus, if n is the order of G, then we have $x^n = e$ for each $x \in G$.

In the case where G is commutative, we often write $x+y$ instead of $m(x, y)$, 0 instead of e, and $-x$ instead of x'; then

$$2x = x + x, \quad 3x = 2x + x, \quad \ldots$$

are defined. A group is often called *additive* when its law of composition is written as $x + y$, and *multiplicative* when it is written as xy.

If G and H are two groups, then a structure is defined on the Cartesian product $G \times H$ by setting

$$(x, y)(x', y') = (xx', yy').$$

We say that the group $G \times H$ thus defined is the *product group* of G and H. This concept can at once be generalized to any number of groups.

C II) Rings

We have already observed that in the set of integers (denoted by Z), or the set of real numbers (denoted by R), or the set of complex numbers (denoted by C) there are *two* laws of composition, addition and multiplication. Problems in the theory of numbers led Euler, and later Gauss in a more systematic manner, to consider subsets of C *other* than R or Z, in which both addition *and* multiplication do not *take one out* of this subset. These subsets are called *rings* (Dedekind called them "orders"). The earliest examples were numbers of the form $a + b\sqrt{D}$, where D is a *fixed* integer not the square of another integer, and a and b are *variable* integers (if $D < 0$, the numbers $a + b\sqrt{D} = a + ib\sqrt{-D}$ are complex). We have in fact

(45) $(a + b\sqrt{D}) + (a' + b'\sqrt{D}) = (a + a') + (b + b')\sqrt{D}$,

(46) $(a + b\sqrt{D})(a' + b'\sqrt{D}) = (aa' + bb'D) + (ab' + ba')\sqrt{D}$.

The ring formed by these numbers is denoted by $\mathsf{Z}[\sqrt{D}]$.

Gauss considered in detail the structure of the ring of the numbers $a + bi$ ($= a + b\sqrt{-1}$), which are now called *Gaussian integers* . He showed that there is an "arithmetic" in this ring very close to normal arithmetic: we can define "prime numbers", and there is a unique way of breaking down each Gaussian integer into prime factors (Appendix 3, §B). It is this factorization which "explains" the theorems of Fermat and Euler on the integer solutions of $x^2 + y^2 = n$ (Chapter IV, §2, A). But further study showed that these arithmetical properties are nor true any more for many values of D (for example $D = -5$). Half a century of effort was needed to overcome this difficulty; and just as the problem of the solution of equations "by radicals" was at the origin of the theory of groups (§1, D), so the study of rings $\mathsf{Z}[\sqrt{D}]$ and of rings generated by "roots of unity" (complex numbers which satisfy an equation $x^n - 1 = 0$) was the source of the great modern theory of *algebraic numbers* .

These examples are not the only ones in which *two* laws of composition can be defined: the set of *functions* with real values defined on an arbitrary set E is also a *ring* with respect to the addition $f + g$ and the multiplication *fg* defined in §1, C); and the same is true of *polynomials* in any number of real or complex variables.

These observations led at the beginning of the twentieth century to the definition of the general structure of *commutative rings*. Here the "basic" objects are a set A and two "laws of composition", denoted respectively by $(x, y) \mapsto x + y$ (*addition*) and by $(x, y) \mapsto xy$ (*multiplication*). The axioms tell us that with respect to addition *alone*, A is an *additive group*, and that multiplication is associative, commutative, and distributive with respect to addition.

C III) Fields

Certain rings (but not all[26]) have a *unit element* e such that $ex = x$ for each x. Furthermore, from the beginning of the nineteenth century, Abel and Galois brought to light (without giving them a name) sets K of complex numbers which are rings containing the set $\mathbf{Q}$ of rational numbers and for which every $x \in K$ such that $x \neq 0$[27] has also its *inverse* $1/x$ *in* K. Since the time of Dedekind, these sets have been known as *commutative fields, subfields* of $\mathbf{C}$. The sets $\mathbf{Q}$, $\mathbf{R}$, and $\mathbf{C}$ are fields. The fields dealt with by Abel and Galois are obtained by taking a finite number of arbitrary complex numbers $a_1, a_2, ..., a_m$: K is the set of numbers $R(a_1, a_2, ..., a_m)$, where $R = P/Q$ is the quotient of two polynomials P, Q of arbitrary degrees in m variables and with rational coefficients, subject to the one condition that $Q(a_1, a_2, ..., a_m) \neq 0$. In fact, for polynomials P, Q, P_1, Q_1 such that $Q(a_1, ..., a_m) \neq 0$ and $Q_1(a_1, ..., a_m) \neq 0$, we have

$$\frac{P}{Q} + \frac{P_1}{Q_1} = \frac{PQ_1 + P_1Q}{QQ_1}, \qquad \frac{P}{Q} \cdot \frac{P_1}{Q_1} = \frac{PP_1}{QQ_1},$$

and the product $Q(a_1, ..., a_m)Q_1(a_1, ..., a_m)$ is not zero. If in addition we have $P(a_1, ..., a_m) \neq 0$, then the number

$$\frac{P(a_1, a_2, ..., a_m)}{Q(a_1, a_2, ..., a_m)}$$

has an inverse

$$\frac{Q(a_1, a_2, ..., a_m)}{P(a_1, a_2, ..., a_m)}.$$

We write $K = \mathbf{Q}(a_1, ..., a_m)$, and we say that K is the field *generated* by $a_1, a_2, ..., a_m$. In this definition, moreover, the field $\mathbf{Q}$ can throughout be replaced by any subfield of the field $\mathbf{C}$ of complex numbers.

---

[26] An example is the subring $2\mathbf{Z}$ of *even* numbers in $\mathbf{Z}$. The product of two even numbers is even, but there is no *even* number e such that $e(2n) = 2n$.

[27] The axioms of rings imply that $0x = x$ for all $x \in A$, for we have

$$x \cdot x = (x + 0) \cdot x = x \cdot x + 0 \cdot x,$$

and since A is a group with respect to addition, it follows that $0 \cdot x = 0$.

We should also remark that the numbers $a + b\sqrt{D}$, where a and b are *rational*, likewise form a field, for we cannot have $a + b\sqrt{D} = 0$, that is to say, $a^2 - b^2 D = 0$, unless $a = b = 0$, because D is not a square. We can thus write, for $a + b\sqrt{D} \neq 0$

$$\frac{1}{a + b\sqrt{D}} = \frac{a}{a^2 - b^2 D} - \frac{b}{a^2 - b^2 D}\sqrt{D},$$

which is also of the form $c + d\sqrt{D}$, with c and d rationals.

C IV) Non-commutative Rings and Fields

The set H of Hamilton's quaternions (§2, D) is "nearly" a field: in fact, if a quaternion

$$q = a + bi + cj + dk$$

is not zero, then we have $a^2 + b^2 + c^2 + d^2 \neq 0$. The formula for multiplication (39) thus shows that the quaternion

$$q' = \frac{a - bi - cj - dk}{a^2 + b^2 + c^2 + d^2}$$

is such that $qq' = q'q = 1$. In other words, every non-zero quaternion has an *inverse* (on the right and the left).

We are thus led to generalize the definition of commutative rings: a *ring* (commutative or not) is a set A in which the following are defined: (i) an addition with respect to which A is a commutative group; (ii) an associative multiplication, *not necessarily commutative* , but *distributive on the right and left* with respect to addition, that is to say, we have

(47) $x(y + z) = xy + xz$ and $(y + z)x = yx + zx$.

A *field* (not necessarily commutative) is a ring K in which the multiplication has a unit element e, and every $x \neq 0$ has an *inverse* x' on the right and left, in other words, such that $xx' = x'x = e$. Usually we write 1 instead of e and x^{-1} instead of x'. It will be noticed that the set K^* of the non-zero elements of K is a *group* with respect to multiplication. The set H of quaternions provides an example of a non-commutative field.

D. Order Structures

We have seen (Chapter III, §4) that the idea of the *order* of points on a line (or the concept of a point "between" two others) is one of the basic geometric concepts. It is a mathematical "model" of the intuitive conception of comparison of "magnitudes" or of integers. In the last third of the nineteenth century, Cantor and Dedekind were led to deepen and generalize this concept in various directions.

At the beginning of his career, Cantor, following Riemann, studied the problems of analysis in which the result could depend *a priori* on a set E of "exceptional" real numbers (Appendix 5, §D). First he was able to show that, if E were finite, then these numbers had no influence on the result. But when E is *infinite*, he could reach the same conclusion only by making use of supplementary hypotheses, which led him to realize what a great variety of sets of real numbers there are if they are considered in the light of the *order structure* of their elements. For example, if E is the set of rational numbers, *between* any two numbers of E there are always others (and even infinitely many others). Consider on the other hand the set E of numbers $1/n$ for $n = 2, 3, ...$: this set does not have the preceding property. Notice too that this set E has a *largest* member $1/2$, but that it has no *smallest* member. The opposite is true for the set F of numbers $1 - 1/n$ for $n = 2, 3, ...$ The union $E \cup F$ has neither a largest nor a smallest member. It is understandable that such examples had the effect of inciting Cantor to undertake a more general study of the idea of order.

Dedekind, for his part, in his researches into the theory of numbers, was led to study the relation "a divides b" between two natural numbers a and b. If we write this in an abridged form $R(a, b)$, we have the following three properties:

1) $R(a, a)$ is always true (this is known as the *reflexivity* of the relation $R(a, b)$);

2) If we have $R(a, b)$ and $R(b, c)$, then we have $R(a, c)$ (*transitivity* of the relation $R(a, b)$);

3) If we have at the same time $R(a, b)$ and $R(b, a)$, then necessarily $a = b$ (a property sometimes styled *antisymmetry*).

Notice that these three properties are obviously also satisfied by the usual relation $a \leq b$ between real numbers which was studied by Cantor. But while we *always* have either $a \leq b$ or $b \leq a$, this is not the case for the relation "a divides b": 3 does not divide 7 and 7 does not divide 3.

These examples and many others led to the general concept of a *relation of order* (or *partial order*) between members of *any* set E: it is a relation $R(a, b)$ between two members of this set which satisfies the properties 1), 2), and 3), above; these are the *axioms of the structure of order*. In the case where in addition $R(a, b)$ satisfies the axiom:

4) For *every* pair of members a, b of E, we have either $R(a, b)$ or $R(b, a)$ (if both are satisfied at the same time, we have $a = b$ by 3));

then we say that $R(a, b)$ is a relation of *total order*.

Contrary to what one might think, we encounter relations of (partial) order far more often in mathematics than those of total order. The one most frequently met is the relation of *inclusion*, $A \subset B$, between the "part" and the "whole", so often used by Euclid, for it satisfies 1), 2), and 3), above, but not 4) as soon as the "whole" has more than two members: if D_1, D_2 are two distinct lines in a plane neither of the two is a subset of the other.

E. Metric Spaces and Topological Concepts

We have seen (Chapter III, §9) that the concepts of *limit* of a sequence of numbers and of *continuity* of a real function of a real variable, which for a long time remained vague, were eventually formulated in a precise way at the beginning of the nineteenth century. These are two of the concepts which are now called "topological".

Other topological concepts appeared after 1860 in the work of Weierstrass and Cantor (as well as in notes made by Dedekind which were not published in his lifetime). Weierstrass, in dealing with the foundations of analysis, defined in his lectures a *neighbourhood* of a real number x_0: it is a subset of the set of real numbers which contains an interval $[x_0 - \alpha, \ x_0 + \alpha]$ of unspecified but non-zero length. Dedekind, in his notes, deduced from this the concept of an *open* set (which at this stage he called a "field"): it is a subset U which is a neighbourhood of *each* of its points. Dedekind even extended this definition to the plane and to space by replacing the interval by a disc or a ball. Cantor for his part, led by his studies of "exceptional" sets (§C, above) to examine the properties of *arbitrary* subsets E of the set $\mathbf{R}$, defined the *derived set* E' of E as the set of points x_0 for which *every* neighbourhood contains an infinite number of points of E (E' is therefore empty if E is finite). He then took the derived set $E'' = (E')'$ of E', and so on, and examined various cases which might arise: for example, for the set of numbers $1/n$ $(n = 2, 3, ...)$, E' is reduced to the one point 0; for the set of numbers of the form $1/n$ and $1 - 1/n$, E' consists of the two numbers 0 and 1 ; in both cases E'' is empty. On the other hand, if E is the set of rational numbers between 0 and 1, then E' is the whole interval $[0, 1]$, and the derived sets which follow are all equal to this interval.

The needs of analysis, however, notably in the calculus of variations[28] led quite early to these "topological" concepts being extended to objects *other* than the real numbers: we need to speak of "limit" or of "neighbourhood" for vectors, functions with real values, curves, surfaces, etc. It was Fréchet who, in 1906, discovered a very simple and altogether general kind of structure which made it possible to give a meaning to "topological" concepts which arise naturally in most problems in analysis: it is what is now known as the structure of a *metric space*. The "basic objects" of such a structure are an *arbitrary* set E and a mapping $(x, y) \mapsto d(x, y)$ from $E \times E$ into $\mathbf{R}$, such that just three axioms are satisfied:

I) $d(x, x) = 0$; if $x \neq y$, then $d(x, y) > 0$;

II) $d(y, x) = d(x, y)$;

III) $d(x, z) \leq d(x, y) + d(y, z)$.

---

[28] This subject, which was introduced at the end of the seventeenth century, was designed to solve the difficult problems of the calculation of the *minimum* or *maximum* of numbers related to *variable* curves or surfaces: for example, to find the closed curve in the plane of given length , not crossing itself, and which encloses the maximum area; or to find, on a surface such as an ellipsoid, curves of minimum length joining two points.

It is clear that in Euclidean space the *distance* between two points x and y satisfies these axioms, for it will be recognized that III) expresses the well-known inequality between sides of a triangle (Euclid, *Elements*, Book I, 20). Thus members of a metric space are generally called *points*, and the number $d(x, y)$ is called a *distance*.

It is altogether remarkable that all "topological" concepts can be defined using *only* the preceding axioms. For example, an infinite sequence (x_n) of points of E has as its *limit* a point y (or "tends to" y) if the sequence of *numbers* $(d(y, x_n))$ tends to 0. If E and F are two metric spaces, a mapping $f : E \to F$ is *continuous* if each time that a sequence (x_n) of points of E tends to a limit y, the sequence of points $(f(x_n))$ tends to $f(y)$ in F. A *neighbourhood* of a point $x_0 \in E$ is a set V such that, for some sufficiently small number $\alpha > 0$, V contains the set of elements $x \in E$ such that $d(x, x_0) \leq \alpha$; etc.

It will be observed that two different distances on the same set can give rise to the same topological concepts: for example, in the plane $\mathbf{R}^2$, we can take as the distance between two points, (x_1, y_1) and (x_2, y_2), the Euclidean distance

$$\sqrt{(x_1 - x_2)^2 + (y_1 - y_2)^2}\,,$$

but also the number

$$|x_1 - x_2| + |y_1 - y_2|\,.$$

We say that on a set E, two distances $d_1(x, y)$ and $d_2(x, y)$ are *equivalent* if the concepts of neighbourhood are the same for these two distances. A set of equivalent distances on E constitutes what is known as a structure of *metrizable space*.

We cannot describe the multitude of questions in which metric spaces of the greatest variety play a part, whether in analysis, differential geometry, algebraic topology or even the theory of numbers. Let us recall here only what are probably the most frequently used spaces other than the spaces $\mathbf{R}^n$. These are the *Hilbert spaces*, lying at the foundation of quantum mechanics (Appendix 4, §C).

F. Superposition and Dissociation of Structures

What we have described in the foregoing sections are the simplest of structures, those which can be called "pure" structures. In the course of the nineteenth century, however, there appeared types of structure which were even more useful for the solution of classical problems. These are those which can be called "mixed" structures, in which there can be several "basic" sets, each provided with a structure, or several structures on the same set. Naturally there are axioms which *connect* these structures to one another. The number of these "mixed" structures is continually increasing: we shall indicate only a few samples.

One of the earliest examples is the structure of a *vector space*, in which there are two "basic" sets, a commutative field K (Section C III) and a commutative group E written additively (Section C I). There is another basic object, a mapping $(\lambda, x) \mapsto \lambda x$ of $K \times E$ into E (λx is also denoted by $\lambda \cdot x$), called *multiplication by a scalar*, which is subject to the following axioms:

$$(48) \qquad\qquad \lambda(x + y) = \lambda x + \lambda y\,;$$

$$(49) \qquad\qquad (\lambda + \mu)x = \lambda x + \mu x\,;$$

$$(50) \qquad\qquad \lambda(\mu x) = (\lambda \mu)x\,;$$

$$(51) \qquad\qquad 1 \cdot x = x\,.$$

The prototypical example is the set $E = \mathbf{R}^n$ of vectors of dimension n (§1, B), with $K = \mathbf{R}$, the multiplication by a scalar being a *homothety* (§1, B and E). The vector space structure on $\mathbf{R}^n$ was worked out in 1843 by Cayley and Grassmann; it is one of the most widespread and most useful in all areas of mathematics.

This structure is generalized in many ways. In the axioms (48) to (51), it is not necessary to suppose that K be a field: it can be replaced by a commutative *ring* A having an identity element[29]. We then have the $A - module$ structure, which is at the centre of the theory of algebraic numbers set up by Dedekind in 1860–70, and of the theory of algebraic functions which Kronecker thought of at the same time.

It is moreover possible to "enrich" a vector space structure by supposing that on E there is not only an an additive group structure, but also a *ring* structure (commutative or not). In this case a new axiom must be added to Axioms (48) to (51) linking the multiplication in the ring E to the multiplication by a scalar, to wit,

$$(52) \qquad\qquad \lambda(xy) = (\lambda x)y = x(\lambda y)\,.$$

We can now say that we have defined an $A - algebra$ structure (in the nineteenth century it was called a "hypercomplex system"). The set of Hamilton quaternions (§2, D) is an $\mathbf{R} - algebra$.

If on the other hand we "impoverish" the vector space structure by replacing K by a *group* G written multiplicatively (and not necessarily commutative), we naturally suppress Axiom (49), since it no longer makes sense, and thus we obtain what is known as an *additive group E with operators* (members of G). We can even suppress all structures on E, and keep only the mapping $(\lambda, x) \mapsto \lambda \cdot x$ of $G \times E$ into E, with only Axioms (50) and (51). We then have a structure of a *set E with a group of operators* G. It is then said that G *acts* (or *operates*) on E, and the mapping $(\lambda, x) \mapsto \lambda \cdot x$ is called the *action* of G on E.

The importance of this last structure arises from the fact than since the time of Klein (1873), it is this which has also been designated as *a geometry*

---

[29] When A has no identity, Axiom (51) is of course suppressed.

of the group G, thus uniting the various geometric concepts (real projective geometry, complex projective geometry, and non-Euclidean geometries) which had been developed since the beginning of the nineteenth century. Classical Euclidean geometry in two or three dimensions is, for Klein, a "geometry" in which E is the plane $\mathbf{R}^2$ or the space $\mathbf{R}^3$, and G is the *group of displacements* (§2, B). A "theorem of geometry" for Klein is a relationship between elements or subsets of E which is *invariant* under the action of G, that is to say, one which still holds when the elements x, y, z, ... of E which arise are replaced by $\lambda \cdot x$, $\lambda \cdot y$, $\lambda \cdot z$, ..., for *each* operator $\lambda \in G$, and the same for the subsets of E. This is obviously what happens for all the theorems of Euclidean geometry.

As an example of a totally different "geometry", let us cite the *half-plane of Poincaré H*, the set of the complex numbers $z = x + iy$ such that $y > 0$. The group G is the *linear group of the plane, preserving orientation*, denoted by $GL^+(2,\mathbf{R})$. It is formed from the *matrices*

$$\begin{pmatrix} \alpha & \beta \\ \gamma & \delta \end{pmatrix}$$

(§2, B) whose members are real numbers and which satisfy the inequality $\alpha\delta - \beta\gamma > 0$. The *action* of G on H is defined by

$$(53) \qquad \begin{pmatrix} \alpha & \beta \\ \gamma & \delta \end{pmatrix} \cdot z = \frac{\alpha z + \beta}{\gamma z + \delta}.$$

(This always makes sense, for we cannot have $\gamma z + \delta = 0$ for any $z \in H$. First, γ and δ cannot both be zero and, second, if $\gamma \neq 0$, we cannot have $z = -\delta/\gamma$ since z is not a real number.) We check immediately that the imaginary part of

$$\frac{\alpha(x+iy) + \beta}{\gamma(x+iy) + \delta}$$

is

$$\frac{(\alpha\delta - \beta\gamma)y}{(\gamma x + \delta)^2 + \gamma^2 y^2}.$$

Since $\alpha\delta - \beta\gamma > 0$, this is > 0 because $y > 0$. We shall come across this "geometry" again in dealing with non-Euclidean geometries (Chapter VI, §1, C). The study of the action on H of certain *subgroups* of $GL^+(2,\mathbf{R})$ is at the root of Poincaré's *Fuchsian functions* (1881).

As an example of a "superposition" of several structures on the same set, let us cite the structure of an *ordered commutative field*. A commutative field structure and a structure of *total order*, denoted by $x \leq y$, are considered at the same time on a set K. The axioms which connect these structures are as follows:

$$(54) \qquad \text{If } x \leq y, \quad \text{then } x + z \leq y + z \text{ for any } z \in K.$$

$$(55) \qquad \text{If } x \leq y \text{ and } z \geq 0, \quad \text{then } xz \leq yz.$$

We recognize here two of the axioms of real numbers with respect to the usual order (Chapter III, Appendix 2), the first of which is essentially one of the "common notions" of Euclid for measures of magnitude. All the *subfields* of the field **R** of real numbers are therefore ordered fields; but there are many more (Appendix 3, §E). There are also commutative fields, like the field **C** of complex numbers, for which there is *no* order structure satisfying the Axioms (54) and (55).

If G is a commutative group written additively, and if we define on G an order structure (total or partial) which satisfies (54), then we say that we have defined on G the structure of an *ordered group*. For example, if $G = \mathbf{R}^2$, we define a structure of a "partially" ordered group by taking as the relation of order

$$(x_1, x_2) \prec (y_1, y_2)$$

the relation

(56) $$x_1 \leq y_1 \quad \text{and} \quad x_2 \leq y_2$$

(where $x_1 \leq y_1$ is the usual relation between real numbers).

We can, however, also define a structure of *total* order on $\mathbf{R}^2$ satisfying (54) by taking instead of (56) the relation

(57) $$x_1 < y_1 \quad \text{or} \quad (x_1 = y_1 \quad \text{and} \quad x_2 \leq y_2).$$

This is what is known as the *lexicographic* order.

Another example of "superimposed" structures is provided by the structure of a *topological metrizable group*. It is defined by taking on the "basic" set G a group structure (commutative or not) and a metrizable space structure satisfying the following axiom:

If two sequences (x_n) and (y_n) of members of G each have a limit,

then the sequence $(x_n y_n^{-1})$ has a limit, and, if $x = \lim x_n$ and

(58) $$y = \lim y_n, \text{then we have } xy^{-1} = \lim(x_n y_n^{-1}).$$

For example, for the group of matrices $GL(2, \mathbf{R})$ defined above, the preceding axiom is satisfied if we define as the *distance* between two matrices

$$\begin{pmatrix} a & b \\ c & d \end{pmatrix} \quad \text{and} \quad \begin{pmatrix} a' & b' \\ c' & d' \end{pmatrix},$$

the number

$$|a' - a| + |b' - b| + |c' - c| + |d' - d|.$$

The first topological groups were considered by S. Lie in 1873, under the name of "continuous groups"; he used them to solve problems of integration of differential equations. They became of major importance in analysis, in differntial geometry, and even in the theory of numbers, in relativity theory, and in quantum physics.

A *metrizable topological ring* A can be defined in the same way by the axiom:

If two sequences (x_n) and (y_n) of A each have a limit, then

the sequences $(x_n - y_n)$ and $(x_n y_n)$ each have a limit,

and, if $x = \lim x_n$ and $y = \lim y_n$, then we have

(59) $x - y = \lim(x_n - y_n)$ and $xy = \lim(x_n y_n)$.

The simplest example is obviously the field $\mathbf{R}$, but these rings appear in all areas of mathematics, even in the theory of numbers: on the ring $\mathbf{Z}$ of integers, which seemed to Gauss the "discontinuous" object par excellence, there exists an *infinity* of distances (one for each prime) for which the distance from 0 to a whole number $n \neq 0$ can become *arbitrarily small* and not zero (Appendix 4, §D). However strange this may seem, these distances have acquired a considerable importance in the theory of numbers.

It can be seen in particular that on the set of real numbers various structures intersect: those of field, of order, and of topology. There is also the concept of *measure* on $\mathbf{R}$, which we cannot describe in detail on account of its technical nature, but which plays an essential part in analysis because it makes it possible to attach a positive number or zero not only to the intervals $[a, b]$ (where this is simply the "length" $b - a$), but also to all the sets which are encountered in major problems. It was essentially Cantor whom we have to thank for being able to see the multiplicity of points of view from which the set of real numbers can be considered: during the greater part of the nineteenth century the term "continuum" was used without distinguishing between the various structures which were defined on $\mathbf{R}$, and this led to curious confusions, as we shall see in Chapter VI.

4. Isomorphisms and Classifications

A. Isomorphisms

Is it possible to *compare* two structures of the same type on two sets E and F; for example, two groups? First, the sets E and F themselves must be compared, without reference to their structures: for example, if E and F are *finite* sets, they will be considered as different if they do not have the same *number* of members. Since Cantor's time, this idea has been applied to *all* sets, thanks to the concept of *bijection* (or *bijective* mapping): this is a mapping $f : E \to F$ which has the property that for *any* member $y \in F$, there exists *one and only one* element $x \in E$ such that $y = f(x)$. For example, $x \mapsto x^3$ is a bijection from $\mathbf{R}$ onto itself, but $x \mapsto x^2$ is not one, because, if

$y < 0$, it cannot be a square, and, if $y > 0$, there are *two* numbers, $x = \sqrt{y}$ and $x = -\sqrt{y}$, such that $x^2 = y$. Cantor said that two sets E and F are *equipotent* if there exists a bijection $f : E \to F$; there is then also a bijection $f^{-1} : F \to E$, namely, that mapping which brings about a correspondence between $y \in F$ and the unique element $x \in E$ such that $f(x) = y$. Two finite sets are equipotent if and only if they have the same number of members, and a finite set cannot be equipotent to an infinite set. But Cantor discovered the remarkable fact that there exist infinite sets E and F which *are not* equipotent (cf. Chapter VI, §3, B).

Let us now suppose that E and F are two *groups*: they will be called *isomorphic* if there exists a bijection $f : E \to F$ such that, *in addition*, the product $f(x)f(y)$ corresponds by f to the product xy of two elements x and y of E: in other words,

$$(60) \qquad\qquad f(xy) = f(x)f(y).$$

Such a bijection f is called an *isomorphism* from E onto F, and its "inverse" $f^{-1} : F \to E$ is then also an isomorphism, for if x' and y' are two members of F and if $x = f^{-1}(x')$ and $y = f^{-1}(y')$, then we have $x' = f(x)$ and $y' = f(y)$, whence

$$x'y' = f(x)f(y) = f(xy)$$

by (60), and it follows that

$$f^{-1}(x'y') = xy = f^{-1}(x')f^{-1}(y').$$

Let us consider, for example, two *cyclic* groups E and E' of the same order n (§3, C I). Thus E (respectively, E') has a member a (respectively, b) such that E (respectively, E') is formed of *distinct* elements e, a, a^2, ..., a^{n-1} (respectively, e', b, b^2, ..., b^{n-1}) with $a^n = e$ (respectively, $b^n = e'$), the neutral element. A *bijection* $f : E \to E'$ is then defined by taking $f(e) = e'$ and $f(a^k) = b^k$ for $1 \leq k \leq n - 1$. Let us show that we do indeed have $f(xy) = f(x)f(y)$. This is obvious if x or y is the neutral element e. Otherwise, we have $x = a^h$ and $y = a^k$ for some $h \leq n - 1$ and $k \leq n - 1$. It follows that $xy = a^{h+k}$ and $f(x)f(y) = b^{h+k}$. If $h + k \leq n - 1$, we do indeed have $f(xy) = f(x)f(y)$ by the definition of f; and, if $h + k \geq n$, we can write $h + k = n + j$ with $0 \leq j \leq n - 2$, whence

$$a^{h+k} = a^n a^j = a^j \quad \text{and} \quad b^{h+k} = b^n b^j = b^j,$$

and so

$$f(xy) = f(a^{h+k}) = f(a^j) = b^j = b^{h+k} = f(x)f(y).$$

We have thus proved the following theorem: *All cyclic groups of the same order are isomorphic*. Thus two groups as different in appearance as the *additive* group of *classes of congruences, modulo n* (§2, C) and the group of circular permutations $(1\ 2\ ...\ n)^k$ (§2, B) are isomorphic.

The great interest of the existence of an isomorphism from one group E onto another group F is that, for any property of elements or of subsets of E which is expressed only with the help of the group axioms, we immediately deduce that the *same* property holds for the elements or subsets of F which correspond to them by the isomorphism, and which are sometimes much less evident.

The existence of an isomorphism, however, is not always so obvious. Take for example the *additive* group $\mathbf{R}$ of real numbers, positive, negative or zero, and the *multiplicative* group $\mathbf{R}_+^*$ of real numbers > 0. It was not until the seventeenth century that the *isomorphisms* of $\mathbf{R}_+^*$ with $\mathbf{R}$ were discovered – the *logarithms* $x \mapsto \log_a x$ of base $a > 1$. These are in fact bijective mappings, and, disregarding the different ways of writing the two groups, the relation

$$(61) \qquad \log_a(xy) = \log_a x + \log_a y$$

is none other than equation (60).

On the other hand, if we consider the subgroup $\mathbf{Q}$ of the additive group $\mathbf{R}$ formed by the *rational* numbers, positive, negative or zero, and the subgroup $\mathbf{Q}_+^*$ of the multiplicative group $\mathbf{R}_+^*$ formed by the *rational* numbers > 0, these two groups *are not isomorphic* . For there exists, for each $x \in \mathbf{Q}$, an element $y \in \mathbf{Q}$ such that $2y = x$. If there were an isomorphism $f : \mathbf{Q} \to \mathbf{Q}_+^*$, then, for each $x' \in \mathbf{Q}_+^*$, there would exist one and only one $x \in \mathbf{Q}$ such that $f(x) = x'$, and, if $2y = x$ and $f(y) = y'$, we would have

$$y'^2 = f(2y) = f(x) = x'$$

by definition. But it has been known since the time of the Pythagoreans that this is not true if $x' = 2$ (see Chapter III, §2).

For each of the structures considered in §3, a concept of isomorphism is defined in the same way. For two rings A and B, an isomorphism f from A onto B is a bijection such that

$$(62) \qquad f(x + y) = f(x) + f(y) \quad \text{and} \quad f(xy) = f(x)f(y)$$

whatever x and y may be in A. For example, let E and F be two *equipotent* sets, and let A and B be the rings formed from the real functions defined on E and F, respectively. Let $\sigma : F \to E$ be a bijection. Then the mapping $u \mapsto u \circ \sigma$, which makes a function $x \mapsto u(x)$ defined on E correspond to the function $y \mapsto u(\sigma(y))$ defined on F, is an isomorphism of A onto B, as is easily ascertained.

For two ordered sets E and F, an isomorphism of E onto F for the structures of order is a bijection $f : E \to F$ such that the relation $x \le y$ is *equivalent* to $f(x) \le f(y)$[30]. For example, the mapping $x \mapsto 2x$ is an isomorphism of the interval $[0, 1]$ onto the interval $[0, 2]$.

---

[30] When it is not assumed that the structures of order on E are total orders, a

If E and E' are two metric spaces (§3, E), and if d and d' are the distances in these two spaces, an isomorphism of metric spaces (which is also known as an *isometry*) from E onto E' is a bijection $f : E \to E'$ such that

$$d'(f(x), f(y)) = d(x, y).$$

One of the great advances of modern analysis was to show that a certain metric space defined in the theory of integration is isomorphic to a Hilbert space.

If we have only $d''(f(x), f(y)) = d(x, y)$, where d'' is a distance equivalent to d' (§3, E), then we say that f is a *homeomorphism* of the metrizable space E onto the metrizable space E'. This amounts to the same thing as saying that f and f^{-1} are continuous.

B. Problems of Classification

Each time a structure is defined which proves useful for the solution of classical problems, we are naturally led to *classify* all the structures which satisfy the same system of axioms Σ. Of course, two isomorphic structures are not distinguished: in other words, the task is to define a collection of structures satisfying Σ, any two of which are not isomorphic to each other, and such that *every* other structure satisfying Σ is isomorphic to one of the structures of this collection.

Here we have an ideal goal which has been attained only in a small number of cases. In 1865, Schering, inspired by the papers left by Gauss, which he had undertaken to edit, had explicitly introduced the structure of a commutative group in classes of quadratic forms with given discriminants (§2, C). Two years later, Kronecker observed that it is possible, by the same methods, to classify *all finite* commutative groups. The result is very simple: these groups are the products of a finite number of *cyclic* groups whose orders are the powers of prime numbers.

It was quickly realized that to classify *finite, non-commutative* groups presented quite a different order of difficulty. Already, the one problem of determining all the groups of a *given order* N to within isomorphism, is very arduous: no one has got past a few thousand for the number N. A recourse was to fall back on certain types of finite groups verifying additional interesting properties: for example, groups whose order is a power of a prime number p have been much studied (these are called *p-groups*). They have remarkable and very restrictive properties, but in spite of this we are very far from being able to describe all of them.

A problem which goes back to Galois is the determination of the *simple* finite, non-commutative groups, which are in a way the opposite of p-groups: they constitute what can be called "basic building-blocks" out of which all the other finite groups can in theory be reconstructed (Appendix 2, §D). The quest to determine *all* of them (to within isomorphism) went on for 150 years. Galois found the smallest, which has sixty members; but it was not until 1983 that a complete enumeration was obtained, subject to verification (now going on) of the proofs, which cover about ten thousand pages!

Of course, a far greater variety still of infinite groups is to be expected, and it is only by imposing supplementary axioms on them that we can hope to classify some of them. One of the earliest successes was a complete classification of certain *Lie groups* (§3, F), also known as *simple* groups (although they are defined differently from finite simple groups). These were all described betweeen 1890 and 1913 by Lie, W. Killing, and E. Cartan.

At about the same time a classification was likewise obtained of certain R - algebras and C - algebras, also called *simple* (§3, F).

As regards fields, we know how to determine completely only *finite* fields, discovered in substance (without the terminology) by Gauss, who published nothing on this subject, and by Galois. The number of elements of these fields is always a power of a prime number, and for each of these powers p^n there is a unique field structure, to within isomorphism. These fields must be commutative. They have recently become very useful in information theory, to construct what are known as "error correcting codes": use is made in this construction of the "geometry on a finite field", in which "algebraic curves" occur which have only a finite number of points!

Many things are known about *algebraic number fields* which come from one or several roots (real or complex) of equations with *rational* coefficients. The study of these began with Gauss, and they have continued to be the subject of profound research right up to our day: it is one of the main branches of present-day number theory. We are still a long way, however, from having penetrated all of its secrets.

It is understandable that when supplementary axioms are added to those of a structure, the "richer" structures which are thus obtained are less numerous. So it is that among *topological fields* (§3, F), those which are *locally compact* [31] are all known. *Ordered fields* again show much diversity; but if the axiom of Archimedes (see Chapter III, §6), is added to their axioms, a complete description of these ordered fields can be obtained: each is a *subfield* of the field R. And if we require in addition that Cauchy's postulate of nested intervals (see Chapter III, §6) be satisfied, the *only* field which satisfies all these axioms becomes the field R itself (to within isomorphism). It is therefore possible to *define* the field R as the only *ordered, Archimedean and complete field* (this last adjective lets us know that Cauchy's criterion is satisfied).

C. The Invention of Functors and of Structures

Faced with their inability to classify all structures of the same type (that is to say, satisfying the same axioms) "to within isomorphism", mathematicians often fall back on a coarser classification, in which structures which are not

---

[31] A set A is *compact* if, from every sequence (x_n) of elements of A a sequence (x_{n_k}) can be extracted which has a limit in A. A space is *locally compact* if every point has a compact neighbourhood.

isomorphic are all brought together in one class. These are often said to be "equivalent", in a weaker sense than isomorphism. One of the oldest methods of obtaining this kind of classification consists in attaching to a structure a *numerical invariant* , that is to say, a number which is the same for two isomorphic structures, but is not enough to characterize a structure to within isomorphism. We shall give an example of this in Chapter VI, §3, B, along with the concepts of *cardinal number* and of *dimension* , which gave rise to some difficult problems before they were clarified.

A modern version of this idea consists in associating with a structure *another* structure which can be of a different type. This association must conform to conditions connected with the concept of *functor* , which we cannot go into here. These conditions entail that two isomorphic structures of the first type have associated with them "functorially" two isomorphic structures of the second type, but not necessarily the other way round. For example, a non-associative "*Lie algebra* " can be associated with a Lie group; or various *groups* may be associated with a metrizable space, among which *homology* groups and *homotopy* groups are the most important.

It can therefore be said that the *invention of functors* is one of the objectives actively pursued today, and numerous examples show that, even if not all of them have the same important consequences, at least we can expect some to lead to great advances[32].

Another more subtle way of studying a set of objects with the same type of structure, or a set of isomorphism classes of such objects, consists in considering a new type of structure *on this set.* One of the oldest examples (1860) was invented by Cayley. Cayley considered all the curves of given *degree* d[33] in three-dimensional space. He showed that, to each of these curves Γ, there corresponds a *vector* $c(\Gamma)$ of large dimension N (§1, B) for which the N components are constrained by a certain number of algebraic relations, in such a way that to each vector satisfying these relations there corresponds a unique skew curve Γ of degree d. For example, in the case of lines ($d = 1$), one can take $N = 5$, and there is just one relation of the second degree between the five components. For $d = 3$, the theory shows that one can take $N = 399$. We say that the set $C(d)$ of all curves of degree d has been provided with the structure of an *algebraic variety* . In the case of lines, Klein made use of this structure to obtain new results in the "geometry" of lines (a subject which was very fashionable in the nineteenth century). Generalizations of Cayley's method have many applications in current algebraic geometry.

Riemann found some much deeper properties in his theory of algebraic plane curves. Using field theory, he associated a "field of algebraic functions"

---

[32] For example, it can be proved that two structures of the same type *are not* isomorphic by proving that a functor associates with them two non-isomorphic structures. This is how it is shown that the sphere and the torus are not homeomorphic because their corresponding homology groups are not isomorphic.

[33] The degree of a curve is the maximum number of points of intersection of the curve with an arbitrary plane (which must not contain the curve in the case where the latter is planar).

with each irreducible algebraic curve, and he defined the "birational equivalence" of two irreducible curves by the fact that their fields of algebraic functions are isomorphic (Appendix 3, §D). In addition, he associated a whole number, positive or zero, its *genus* [34] g, with every field of algebraic functions on an algebraic curve, in such a way that two birationally equivalent curves have the same genus. Conversely, however, two birationally equivalent curves of algebraic functions with the same genus are not necessarily isomorphic: Riemann showed that the isomorphism *classes* of fields with genus $g > 2$ depend on $3g - 3$ complex parameters, which he called "moduli" [35]. In modern times it has become possible to define a well-determined structure of an algebraic variety on the set of these classes: the study of these is being actively pursued, and it dominates the theory of algebraic curves.

Since then this idea has "colonized" many other areas: sets of mappings subject to certain conditions, sets of isomorphism classes of topological or differential structures, are "organized" by structures of metrizable spaces, of groups, of rings, etc. The *invention of structures* on sets which at first seemed to have none, is another important source of progress.

A concept connected with this idea is that of "genericity" in the classification of structures. It can happen that structures of a certain type tend to present "pathological" aspects so numerous and varied that they discourage all attempts at classification. But it is sometimes possible to remedy this situation by choosing among structures of a given type some which are considered "good", and which can be classified in a reasonable manner. In order, however, to obtain significant results in this way, it is necessary to show also that the "bad" structures which have been eliminated form, in some sense, a "small" subset of the set of all structures of the type being studied. Of course, this idea must be made precise, which can be done in several ways: for example, in an infinite set, the finite subsets can be considered "small", and in a set equipped with a "measure", the sets of measure zero are "small", etc.

5. Mathematics of Our Day

A. A Panorama of Mathematics

To categorize the innumerable works which are published every year in mathematics is a superhuman task, on account of the close-knit network, almost

---

[34] If a plane curve is of degree n, and has no other singular points than the double points with distinct tangents, its genus is $(n - 1)(n - 2)/2 - d$, where d is the number of double points. A curve with equation

$$ax^n + by^n - c = 0,$$

where $abc \neq 0$, has no singular point, and is thus of genus $(n - 1)(n - 2)/2$.

[35] We have already met this word with other meanings: it is a regrettable habit of mathematicians to give the same name to very different objects.

impossible to unravel, of links which bind together the various theories, and which are the mainsprings of their progress. The list which follows, then, must not be seen as anything else than a very approximate indication of the various directions in which mathematical research might be taken, with respect to problems which often do not seem to point in these directions.

I) *Formal logic and set theory*

Although Leibniz (in writings which for a long time remained unpublished) may seem to be a precursor, these subjects did not begin to be broached by mathematicians until about the middle of the nineteenth century. We shall come back in Chapter VI, when we deal with the problems of "foundations", to their origin, to their considerable development during the twentieth century, and to their current position with respect to other areas of mathematics. It must be remarked that formal logic has recently discovered a completely unexpected field of application in computing, in the theory of "automata" of various types, and in everything that can be grouped under the heading of "information theory".

II) *Combinatorics*

To start with, it can be said that what was designated by this name was a part of set theory, namely the theory of *finite* sets, and the problems of estimating the *numbers* of members of such sets, constructed by following various procedures. Some of these problems go back to classical times, for instance, the calculation of the number of pairs of two elements of a set of n members, equal to $n(n - 1)/2$. We met another example of this in the calculation of the number $n!$ of permutations of a set of n members (§1, D). The methods which belong to this study were for a long time rather despised by mathematicians, who were inclined to categorize them as "mathematical recreations", of the kind which amuse many readers of popularizing scientific publications. But within the last few decades the situation has changed considerably: it has been noticed that "combinatoric"methods could have important applications in entirely "respectable" areas of mathematics, such as group theory, Lie algebra theory, algebraic geometry, and algebraic topology. They have also been used successfully in various applications of mathematics, including "information theory".

III) *Categories and functors*

This is one of the newest areas of mathematics, born in 1943 from the reflections of Eilenberg and MacLane on algebraic topology. It is not possible to speak here of what can be considered as a "second degree" abstraction, in which sets equipped with structures and mappings between these sets disappear, or rather are "sublimated" into "objects" and "arrows" devoid of any connection with everyday notions of "collection" or "law". This theory was first used in algebraic topology, and then became of basic importance in algebraic geometry. It reached many other areas of mathematics, and in the end became the object of autonomous development.

IV) *Algebra*

General concepts about the basic algebraic structures: groups, rings, fields, and modules, are now part of the instruction in algebra given at the level of the first years at a university. The greater part of current research in algebra is centred on more specialized algebraic structures. These were initially considered as simple examples which had not been developed very far, but methods and results in this field have proliferated to such an extent that the study of each of them has become a major theory which has to be considered separately. These are listed below in items V) to VII). Under the general heading "algebra" can be left what is known as "universal algebra", in which the laws of composition are not subject to such restrictive axioms as those which define the structures of groups or rings.

As for "classical" algebra, that is to say, essentially the study of systems of algebraic equations, that is now part of commutative algebra (item VII).

V) *"Abstract" groups*

This is a theory which, beginning with Jordan's *Treatise* (1870), has acquired its own methods and problems and has become enormous. It is also undoubtedly the one which has the largest number of uses in all areas of mathematics; to such an extent that it has been said that when you do not quite understand the properties of new mathematical objects, you should try to put a group structure on them. This seems like a whim, but in fact it has more than once succeeded. The areas of mathematics which are closest to "abstract" groups are Lie groups, algebraic geometry, algebraic and differential topology, and number theory; using them as intermediaries, group theory fertilizes many other questions in mathematics and its applications to physics.

VI) *Associative non-commutative algebra and non-associative algebra*

These are some theories which go back to the middle of the nineteenth century. They are concerned with associative, non-commutative *rings* (such as quaternion fields (§2, D)) or with non-associative rings (in which the multiplication is not associative, but remains distributive on the right and left with respect to addition). Attempts are made to *classify* them, usually by imposing supplementary axioms. They come into "abstract" group theory, into Lie group theory, into algebraic topology, and into functional analysis.

VII) *Commutative algebra*

This is the study of commutative rings and fields: it began about 1860, and has also become a separate branch of algebra. It too has its own problems and methods, and, most importantly, it now has a close symbiotic relationship with algebraic geometry. It provides the latter with its technical tools, the properties of commutative rings; but recently it has in its turn benefited from intuitions which belong naturally to problems of algebraic geometry, since it is now possible to *translate* problems of commutative algebra into geometric terms in what is known as Grothendieck's theory of *schemes*, the ultimate generalization of the concepts of algebraic curve and algebraic surface.

VIII) *Homological algebra*

This is concerned with constructions of *functors*, which associate with objects of every category other objects which are most often commutative groups. It is a theory which had its birth in about 1943, at the same time as the theory of categories and of functors (item III), and in connection with algebraic topology. Its field of application has, however, been progressively extended into almost the whole of mathematics. Unfortunately it is impossible to say more about it here, because of the extremely abstract nature of the concepts which it uses.

IX) *General topology*

The beginning of this area of mathematics can be placed in 1906, with the definition of metric spaces (§3, E). As with algebra, the axioms and general properties of the most frequently used topological structures (metrizable spaces and their simplest generalizations) are now subjects of instruction in the first years of a university course. Here too the major theories of algebraic and differential topology (item X) and of function spaces (item XIII) rapidly became detached from these general considerations. Research is being done on topological structures which are hardly used in other areas of mathematics.

X) *Algebraic and differential topology*

To classify all topological structures to within isomorphism is a chimerical undertaking. Endeavours quickly became restricted to those spaces which are closest to our geometric intuition, such as surfaces and the "finite-dimensional varieties" which generalize them and which are indispensable in functional analysis (items XIII to XVI) and in its applications in physics and mechanics[36]. Then ways were found of extending this area by various constructions of a geometric nature. For all these spaces it is possible, as we said above (item VIII), to define *functors* associating an algebraic structure with a space, and algebraic and differential topology are the study of these functors. Created in 1900, it has been developing explosively for fifty years. It is perhaps the area of mathematics in which the greatest number of entirely new ideas has appeared, many of which have had unexpected repercussions in theories which seem very remote from it. It forms an imposing edifice, constantly under renovation, and of such complexity that very few specialists are capable of encompassing all of it.

---

[36] For example, the position of a solid body in space depends on six parameters, and can therefore be marked by a "point" of a six-dimensional space. To see this, consider for instance the position of a cube in space: first its centre, which is given by its three coordinates, must be fixed; once that is done, the perpendicular which passes through the centre to one of the sides (a "quaternary axis" of rotation, cf. Appendix 2, §E) depends on two parameters which can be made concrete as the coordinates of the point where this axis pierces a sphere of radius 1 whose centre is the centre of the cube. Finally, once this axis is fixed, it is still possible to make the cube rotate in any way around it, which gives one more parameter, the angle of rotation.

XI) *Classical analysis*

I have already alluded (§1, A) – for want of a more precise description – to one of the finest blooms of nineteenth-century mathematics, the great theory of analytic functions of *one* complex variable, developed by Cauchy, Riemann and Weierstrass. It continues to this day to attract numerous mathematicians, and it benefits from techniques borrowed from algebraic topology and functional analysis.

In the last third of the nineteenth century, following on the works of Cantor, a branch of classical analysis was also developed, in which analytic functions are deliberately left aside; the aim this time being to see what can still be said of a positive nature about much more "irregular" functions (cf. Chapter VI, §2, B). Here we are concerned with functions of any number of real variables, and this research is now on the whole considered as dependent on topology.

More than any other area of mathematics, classical analysis has branched out in many new directions, the most important of which are described in items XII to XIX below.

XII) *Integration and probability theory*

We have seen (Chapter III, §9) that the seventeenth century saw the definition of an integral on an interval of a real continuous function of a real variable. In the eighteenth century the definition appeared likewise of the *double integral* of a real function $f(x,y)$ of two real variables, defined on a subset D of $\mathbf{R}^2$: if $f(x,y) \geq 0$, the integral

$$\iint_D f(x,y)\,dx\,dy$$

is the volume defined by the relations

$$(x,y) \in D, \quad 0 \leq z \leq f(x,y)$$

in $\mathbf{R}^3$ (for example a hemisphere, if we take for D the disc

$$x^2 + y^2 \leq R^2$$

and $f(x,y) = \sqrt{R^2 - x^2 - y^2}$). It was just as possible in the eighteenth century to define "multiple integrals" of a real function of any number of real variables, although there is no longer any geometric visualization in this case.

From 1895 to 1930 it came to be realized that the concept of integral can be extended much further still. It can now be applied to very general real functions defined on a set E which can be anything, provided that it is possible to define a *measure* for certain subsets A of E, that is to say, to attach to each of these subsets a number $\mu(A) \geq 0$, which generalizes the classical concepts of length, area and volume, and is subject to a small number of axioms. One of these states that for a union $A \cup B$ of two sets *with no common member*, we have

$$\mu(A \cup B) = \mu(A) + \mu(B) :$$

the measure is *additive*.

This ability to define measures and integrals in much wider contexts than classical analysis has brought about huge advances, notably in functional analysis and in Lie group theory. In addition, it has given new life to the calculus of probabilities. Founded in the seventeenth century by Pascal and Fermat, this science remained almost entirely confined to the theory of games of chance; despite notable advances due to James Bernoulli and to Laplace, it lacked solid foundations, and attempts to extend it to what were called "geometric probabilities" did not meet with much success[37]. In 1930, Kolmogorov showed that it is possible to lay the foundations for a theory of probability by attaching it to measure theory: all that is necessary is to consider only those measures which have values between 0 and 1[38]. Definitively incorporated into mathematics in this way, probablity theory has become an area which is expanding at full speed, and its innumerable applications to other sciences are too well known to need insisting on here.

XIII) *Function spaces and operators*

From the seventeenth century, the problems of geometry, of mechanics, and of astronomy (for example, Newton's deduction of Kepler's laws from the law of universal attraction) led to "functional" equations, that is to say, ones in which the "unknowns " were no longer numbers, but functions. We shall speak later (items XV and XVI) of the earliest types of these equations, differential equations and partial differential equations.

At the beginning of the twentieth century, with the idea of metric space (§3, E), it became normal to think that, just as the solution of a system of algebraic equations with n unknown quantities can be considered as a *point* (or a *vector*) in $\mathbf{R}^n$, so a function which is the solution of a functional equation is a *point* in a metric space. To be able to make use of this language, we have to say what the *distance* (§3, E) between two functions is: it soon becomes clear that there are many different ways of defining such a concept (Appendix 4).

Once such "function spaces" have thus been defined, the next step is to consider some *mappings* (§3, B) from a function space E into a space F: these are what are often called *operators*. In the same way as real functions of real variables, they can be put together in various ways (§1, C), and can therefore form *groups* and *rings*; and it is also possible to define integrals relating to operators, etc. Classical physics had already been able to make use of these "abstract" theories (item XIV); but they became of preponderant importance with quantum mechanics, in which physical properties (position, momentum, energy, etc.) are interpreted as *operators* on a Hilbert space.

---

[37] According to their conception of what such a probability was, several authors might give different values to the probability of the same event.

[38] The measure of the whole set must be equal to 1, that of the empty set to 0.

We must also mention the theory, now developing apace, of *optimal control*, the modern form of the calculus of variations, in which everything takes place in function spaces.

XIV) *Commutative harmonic analysis*

The origin of this theory is to be sought in the Pythagoreans' discovery of the "decomposition" of a sound produced by the vibration of a taut string into the "harmonics" of a "fundamental sound". The mathematical theory originated in the eighteenth century: it consisted in the "decomposition" of a *periodic* function f (that is to say, one such that $f(x + 2\pi) = f(x)$ for every $x \in \mathbf{R}$) into the *sum* of "simple" periodic functions

$$a_n \cos(nx + \theta_n).$$

This decomposition must, in general, involve an infinite number of terms, or in other words, it is a *series*. Its study in depth began with Fourier in 1807, and since then it has never stopped attracting analysts, and has had innumerable applications in physics (Appendix 5).

Alongside these "Fourier series", Fourier had extended the scope of his study by creating what is known as the "Fourier transform", one of the earliest examples of an *operator* (item XIII) which set up a correspondence between a real function defined on $\mathbf{R}$ (but not in general periodic) and another function also defined on $\mathbf{R}$. This operators and its generalizations have shown themselves to be tools of the highest utility in the theory of partial differential equations (item XVI), in mathematical physics, and in the calculus of probabilities. Much more recently it has been noticed that there is a close link between this theory and the theory of metrizable commutative groups (§3, F), and harmonic analysis has conquered algebraic number theory (item XXI).

XV) *Differential equations*

Apart from a tiny number of cases, known since the eighteenth century, it is not known how to write down explicitly the solutions of a differential equation[39]. With the exception of equations arising from problems of the vibration of strings or of rigid rods, the general theory of systems of differential equations had for a long time been limited to a *local* study of solutions, according to the nature of the equation in the neighbourhood of a point. It was H. Poincaré who, beginning in 1880, was the first to succeed in obtaining theorems on the *global* bearing of the solutions, which are often called *trajectories* because of the assimilation of the independent variable t to "time". This theory is also nowadays designated the theory of *dynamical systems*, and of course it has multiple applications in mechanics and in astronomy. It has expanded enormously in the past century, and has developed closer and closer links with algebraic topology (already used by Poincaré) and with measure theory, in problems known as "ergodic", arising from physics.

---

[39] Liouville was able to prove that the solutions of equations such as $xy' = e^x$, in which only "elementary" functions appear, are not themselves "elementary" functions.

XVI) *Equations involving partial derivatives*

The first partial differential equations appeared in the eighteenth century, in connection with problems of differential geometry (item XVII) and of physics; but, for lack of appropriate mathematical tools, it was not possible to begin work on the general theory of these until the nineteenth century. At first it was thought that the methods used for the study of differential equations could be easily generalized when passing from one to several independent variables. But the realization soon had to come that the analogies between the two theories were very superficial; the problems posed by partial differential equations called for completely new methods of attack, and studying them has sometimes led to surprising results, such as the existence of equations which have no solution, discovered in 1956. Since 1950, the new concepts drawn from functional analysis and from Fourier transform theory (item XIV) have made possible astonishing new departures in the theory of partial differential equations, which is now expanding at full speed and is still closely linked with applications to physics.

XVII) *Differential geometry*

Since the seventeenth century, mathematicians had applied the differential calculus to the study of *local* properties of plane curves, such as the determination of the tangent at a point, points of inflection, multiple points, and the definition of curvature. These results were extended in the eighteenth century to skew curves and to surfaces. Then Riemann, in 1854, had the temerity to conceptualize "varieties" of any dimension, and to define for them differential concepts which generalize the earlier notions, even though they can obviously no longer be given any "concrete image", as can curves or surfaces. Riemann's successors have considerably developed his ideas, and launched into *global* problems, in ever closer partnership with algebraic topology. This work, which seems to be very far removed from natural science, has proved to have an altogether unexpected application to it in the theory of general relativity and in cosmological theories in which the idea of a space of four or more dimensions is fundamental.

XVIII) *Analytic geometry*

When, in the latter years of the nineteenth century, mathematicians wished to extend to analytic functions of several complex variables the deep results obtained by Cauchy, Riemann and Weierstrass on the functions of a single variable, they came up against unexpected difficulties. It was not until towards the middle of the twentieth century that these were overcome, thanks in part to the use of new concepts emanating from algebraic topology ("sheaf cohomology"). From that moment, the theory has also taken a more geometric turn, bringing it close to algebraic geometry. It has become necessary to study "analytic varieties" alongside "algebraic varieties": for example, an equation

$$F(x, y) = 0$$

between two complex variables, where F is no longer a polynomial but an analytic function, defines an "analytic curve". Thus the theory of functions of several complex variables is now called *analytic geometry*[40].

XIX) *Lie groups*

The metrizable groups introduced by Lie under the name of "continuous groups" (§3, F) for a long time held a marginal position among all the other mathematical theories. Thanks to the labours of E. Cartan and H. Weyl between 1910 and 1930, mathematicians became aware of their importance in differential geometry. Then the area in which they could be of use was progressively enlarged, until nowadays it takes in the whole of mathematics, from the theory of "abstract" groups and the theory of numbers right up to mathematical models of the physics of "elementary particles"; the theory of automorphic functions, originated by H. Poincaré, occupies a central position in this subject.

XX) *Algebraic geometry*

Although the Greeks had studied certain algebraic curves and surfaces, the general study of these objects could not begin until after the invention of the method of coordinates by Descartes and Fermat. From the nineteenth century onwards, the introduction of complex values and of "projective" concepts has given new life to ideas less dependent on the use of coordinates. These ideas have met with remarkable success, and have made the theory very popular among mathematicians. At the same time Riemann, through his deep results on algebraic functions, brought geometry back into contact with the theory of analytic functions of complex variables and with topology. These links are more productive than ever, but from about 1930, it has been possible to replace almost entirely the proofs which made use of analysis with others which depend only on commutative algebra. This has brought about a great expansion of the theory, and has brought it close to the theory of numbers, creating suggestions for problems which are just as useful in the one as in the other of these two theories.

XXI) *Theory of numbers*

This is undoubtedly the oldest of theories, and for many mathematicians it is still the "queen of mathematics", to use Gauss's phrase. Problems in it have always fascinated researchers, because of the apparent simplicity with which they are stated, and the immense difficulties which must be surmounted in order to solve them (see Chapter IV). It has always successfully drawn on algebra and analysis, and now draws on topology too. Reputedly the "purest" of mathematical disciplines, it has nevertheless recently begun to have some applications.

---

[40] From the end of the eighteenth century, the term "analytic geometry" was used for the application to Euclidean geometry of the method of coordinates. This nomenclature is no longer used in our day outside elementary schooling.

B. Specialists and Generalists

A young mathematician who has acquired the indispensable basic knowledge in three or four years of study at a university, and who wishes to specialize in one of the twenty-one items listed above, must still read many books and voluminous memoirs to get abreast of the techniques and problems in the area which he has chosen. It is therefore not at all surprising that most mathematicians confine themselves throughout their careers to one of these theories, even to a single one of its multiple aspects. Some of these, by extending known techniques beyond what their predecessors could do, manage to give these techniques enough power to overcome obstacles which defeated the latter: they may be compared to tacticians who by a fierce fight conquer a strongly defended fortress. But there are also a few minds able to master several areas which seem far removed from each other, and to see ways of using results in one to solve problems in another, like strategists who use clever devices to raze almost without effort positions which had been reputed impregnable. One of the most recent examples is typical in this respect: by using the properties of certain partial differential equations stemming directly from the physics of elementary particles, a young English mathematician, S. Donaldson, has found answers to questions about the topology of four-dimensional differential varieties which had baffled specialists, who would never have imagined that it was possible to tackle them in this way.

Nearly every year for the past forty years a new arrival has thus succeeded in unblocking problems outstanding for decades. There is no reason to think that this continuous flow of talents is fated to dry up: mathematics has never flourished at any epoch as it does in the present.

C. The Evolution of Mathematical Theories

It is well known that in the natural sciences, observations and experiments are organized into explanatory systems ("theories") in which, from a small number of entities more or less open to observation, and from laws which are supposed to describe their behaviour (for example, Newton's laws for the mechanics of "material points"), we are to *deduce* logically what will happen to the phenomena being observed. These correspondences between theory and observation have always been tainted with errors which can be attributed to the imprecision of the instruments of measurement, so long as they are not too serious. But when they become so great that the theory can no longer explain them, the theory must be modified, if necessary by completely overturning its foundations. As is well known, this has been done several times during the last eighty years.

There is nothing like this in mathematics. Once the proof of a theorem from a system of axioms has been recognized as correct, the theorem is never again brought into question: Euclid's theorems are as valid today as they were 2300 years ago. Does this mean that there is no evolution in mathematics? Everything that we have described in this and the preceding chapters proves the contrary. This evolution, however, does not consist solely in the accumulation of new theorems; for these are not simply superimposed on the old ones, but they absorb them, transforming them into "corollaries" which in the end are sometimes not even explicitly mentioned any more, unless by historians. Even the formulation of theorems can change completely: for example, the description of regular polyhedrons, the culminating point of Greek mathematics, is now expressed as the description of the finite subgroups of the group of displacements of three-dimensional space (§2, B). Other transformations may appear even stranger: a system of axioms is replaced by an "equivalent" system[41], in which a statement which was a theorem in the old system becomes an axiom in the new one, while the old axioms become theorems[42].

The change in the conception of mathematics which takes place between one generation and another is therefore a *reorganization*, which takes account of new acquisitions and of their relations to older theorems. Ever since Euclid, this change has always been consolidated in expository works; since the beginning of the nineteenth century, these works have been chiefly text-books intended for students at different levels. It is by comparing these works that historians of science can learn the ways of thinking of members of the mathematical community at a given time, and not only of the geniuses who dominated the epoch. Almost always the thought of the latter went far beyond that of their contemporaries, and prefigured what would come to be commonly accepted in the succeeding generation. Nowadays mathematical theories are expounded by sorting them into rough categories like those itemized above, according to the structures which characterize them. But it has been done in this way only for about fifty years, while historically speaking the conception of these structures and of how to use them goes back to the nineteenth and to the first third of the twentieth century, as we have tried to show.

If mathematics has continued to progress ever more rapidly since the sixteenth century it is above all thanks to the uninterrupted flow of problems which mathematicians have put to themselves. "A branch of science", said Hilbert, "is living so long as it offers a host of problems. A lack of problems means that it is dead, or has stopped developing."

For most of the items listed in Section A), this indispensable supply has never dried up since the theory first existed, whether it is a question of problems posed by the applications of mathematics, or of problems arising from an "internal" evolution like those which we gave as examples in Chapter IV.

There are nevertheless some exceptions. We pointed out at the beginning of this chapter the curious hiatus of 1785 – 1795 in the progress of analysis, perhaps caused by a feeling of powerlessness to deal with the problems in physics of that time, for want of the invention of new methods.

Another example of a brilliant flourishing followed by a decline is the *theory of invariants*, on the borders of algebra and algebraic geometry. After

---

[41] Two systems of axioms are equivalent if all the theorems deduced from the one can also be deduced from the other.

[42] Some mathematicians of conservative temperament never fail to protest against such practices

being the subject of a multitude of works for half a century (1840–1890), the most important problems raised in it were all solved by Hilbert, and almost immediately afterwards, for lack of new problems, the theory became dormant for fifty years. It is only recently that an upsurge in activity has become manifest, on account of its links with Lie group theory (item XIX) and with the theory of Riemann's "moduli" in algebraic geometry (§4, C).

We have many times emphasized the advances which can result from the *convergence* on a single problem of apparently very different theories. But instances of *divergence* can also occur. For example, we said that general topology (item IX) came into being in order to provide structures which would make it possible to deal more efficaciously with problems in functional analysis. It is possible, as Hamilton did (§2, D), to *vary* the axioms of these structures, and to study the properties of the new structures which are obtained in this way. Except in rare cases, however, they are of no further use in the study of operators or of functional equations, so that the end result is two areas of mathematics which are no longer in contact. The same phenomenon occurs in algebra, for which the "consumer" theories are chiefly algebraic geometry, Lie group theory, and the theory of numbers: many studies belonging to items VI and VII have a tendency to split off to the extent of becoming almost totally separate. Not everybody has the (posthumous) luck of Hamilton.

6. Intuition and Structures

All great mathematicians who have spoken about their work have been happy to insist on the part played in it by what they generally call their "intuition"[43]. This may seem strange to the non-initiate: if he opens a book on mathematics today, he will see nothing in it but hundreds of lemmas, formulae, theorems, and corollaries, linked together in a complicated way according to implacable rules of logic, and relating to mathematical objects which cannot be "pictured" in our physical universe. I have known older mathematicians, undisputed masters of classical analysis, who could not conceive how their juniors found their way unhesitatingly in a sea of "abstractions". They would readily have likened their modes of reasoning to the work of machines, manipulating formulae without attempting to understand them.

I believe that nothing is further from the truth; but obviously we must not take the word "intuition" in the sense normally given to it. The difficulty is that what a mathematician calls "intuition" is for him an entirely personal psychological experience, scarcely communicable; and there is every reason to think that the "intuitions" of two mathematicians are often very different.

Perhaps, however, it is not an altogether illusory aim to point to certain features of this experience which I think are fairly general, even though based chiefly on my own recollections. In the first place, when you begin to be

---

[43] English-speaking people are more likely to use the word "insight", which is perhaps more appropriate than a term which has a whiff of the "magical".

interested in a theory which you did not know before, you have *no* intuition, even though you can verify step by step all the theorems belonging to the theory. You put questions which later will seem stupid, and you are completely incapable of thinking on your own of arguments analogous to those you have read. Then, if you persevere, little by little the veil is lifted; you begin to understand why the mathematicians who contributed to the theory proceeded in one way rather than in another. The objects treated by the theory become familiar; you realize that they have "natural" ways of behaving which you must be careful not to interfere with[44]. It is at this moment that you may have the luck to light upon a new theorem or a new method of proof going beyond the preceding ones. "Intuition" has come, but intuition is not enough: the proof which you have glimpsed must be subjected to the inexorable rules of logic, which must govern all its parts: a painful labour, which at times is sure to meet unexpected and demoralizing set-backs.

How can the mathematician of today engage in this quest for new discoveries – which has been no doubt the same in all epochs, as we have seen, for example, in the case of Hamilton (§2, D) – when the concepts with which he deals are entirely devoid of any "image" accessible to the senses? I believe that he creates for himself purely mental and incommunicable images of these mathematical objects. The precise formulation of the axioms which define them may help in the formation of these images, by eliminating all the unnecessary particular features which may be attached to them in the various uses of their structures: in other words, although this may seem paradoxical, abstraction may be useful for the forming of "intuition" rather than paralysing it.

Even more fruitful, it seems to me, are what can be called *transfers* of intuition. As we have already said several times, a single structure (for example, that of a vector space (§3, F)) is to be found in a number of theorems. In each of these, this structure is connected with particular features of the theory which are not found in another. It can happen that in modifying the language of this latter to fit the present need, one can also bring into it particular features of the former, and this can sometimes give rise to new and very fertile "intuitions".

The archetypal example of these transfers is the diffusion throughout all mathematics of *the language of geometry*. After the invention, by Descartes and Fermat, of the method of coordinates (see Chapter III, §8), which reduced geometric problems to problems in algebra, it was not long before it was noticed that, conversely, algebraic problems which did not entail more than two or three variables could be interpreted as equivalent to problems in geometry, if the variables were considered as the coordinates of a point. This transcription gave a stronger grasp to an intuition drawn from experience in geometry, an intuition which could serve as a guide to solving the problems.

---

[44] For example, when two structures are isomorphic (§4, A), it may be that there exist some isomorphisms from one onto the other which are more "natural" than others. It was an attempt to make precise this somewhat vague idea which led to the concept of "functor".

Unfortunately, this geometric interpretation disappeared as soon as the number of variables went beyond three. Now from the eighteenth century onwards, notably in mechanics and in astronomy, this number (that of the "degrees of freedom" of the system being studied) could be arbitrarily large. From the middle of the nineteenth century, mathematicians decided to treat these problems using a *conventional* language, copied from that of classical geometry, although no longer related to experiential reality: thus we speak of a "point in n-dimensional space" instead of a system of n numbers (§1, B), of a "hyperplane" to designate the set of "points" $(u_1, u_2, ..., u_n)$ satisfying a linear equation

$$a_1 u_1 + a_2 u_2 + ... + a_n u_n + b = 0 ,$$

where the coefficients are real and the a_j $(1 \leq j \leq n)$ are not all zero; and so on. In the problems posed by mechanics, the variables are moreover linked in general by relationships; thus we have to consider only the "points" $(u_1, u_2, ..., u_n)$ situated on the subsets which satisfy these relationships, sets which are usually termed "varieties": if there are p relationships, the "variety" is said in general to be "of $n - p$ dimensions"[45].

Thus, from the latter years of the nineteenth century, the studies of algebraic geometry and of differential geometry, formerly confined to ordinary space in three dimensions, were enlarged to take in varieties in any number of dimensions; and since the time of H. Poincaré, it has been considered that varieties constitute the area in which mathematical analysis must develop (it is also termed "global analysis"). Experience proves that a good number of results in three-dimensional geometry have "natural" generalizations in this enlarged frame. To give a simple and somewhat crude example: the fact that in ordinary space a line and a plane meet "in general" at a single point has as its extension the fact that in n-dimensional space we must expect a variety of p dimensions and a variety of $n - p$ dimensions to have an intersection formed "in general" of isolated points. This transfer of the usual "geometric intuition" to a new "abstract intuition" which lets itself be guided by verbal analogies has been shown to be of remarkable efficacy, and it would be a pity not to make use of it. Almost always the replacement of algebraic language by geometric language makes for considerable simplification and brings to light properties which remain unsuspected when they are buried in a confused heap of calculations.

So it is that mathematical models of linear programming and of optimization are understood and dealt with much better when the inequalities which arise in them are interpreted in terms of the properties of convex sets in a multi-dimensional space. The classification of semisimple Lie groups, which at first was done with a barrage of calculations of determinants, is now based on the study of geometric configurations, "Weyl chambers", whose marvellous properties make the most elegant results of classical geometry look pale.

---

[45] Of course this language has not changed any of the substance of the theorems, which must always be proved by algebraic calculations: the new language contributes at most briefer and more expressive statements.

Starting from these configurations yet other objects with even more astonishing properties have recently been invented: the "buildings of Tits", source of fruitful applications and one which in all likelihood would never have been conceived without geometric language.

It can therefore easily be understood that mathematicians have been so appreciative of the advantages of this language that it has not taken them long to go well beyond the bounds of these first generalizations. Classical linear algebra manipulated systems of numbers (§1, B) and matrices (§2, B): in geometric language these become vectors and linear maps. Now, however, this language is extended to the point where "scalars", instead of being numbers, are members of any field or even of any ring (§3, C II), which has the effect of making the arguments appear simpler; so that nowadays we cannot even see how it would be possible to do without this "spatial" language, when for instance we confront the most recent generalization of linear algebra, homological algebra (§5, A, VIII).

Classical algebraic geometry is generalized in the same way by introducing as objects of study varieties defined by equations with coefficients in any field, while keeping the language inherited from the case where this field was the field of real numbers or complex numbers. This was possible whenever proofs could be constructed without using analysis.

Geometric language was also generalized in quite another direction. We have explained (§5, A, XIII) how functions came to be spoken of as "points" in a function space. This geometric language led to unexpected and fruitful connections. For example, a difficult classical problem, "Dirichlet's problem" of potential theory, is interpreted as the generalization, in an appropriate "Hilbert space", of the classical geometric operation of orthogonal projection onto a line or a plane. But that is not the end of this new bursting forth of "topological" concepts: they have invaded even the theory of numbers, in which we can now speak of "local" properties as if we were putting ourselves in the neighbourhood of a point. Thus we see that it can now be unhesitatingly affirmed that the geometric spirit is more than ever dominant over the whole body of mathematics.

We have emphasized this "transfer" of ideas drawn from that bias toward geometric reasoning acquired with our basic schooling, because it is undoubtedly the most striking and the most extensive. There are, however, many analogous transfers rising from algebra, or from analysis, or from the theory of numbers. They are indicative of the dominant feature of current mathematics, which cannot be too much emphasized, its *unity*.

V Appendix

1. The Resolution of Quartic Equations

If x_1, x_2, x_3, x_4 are the roots of the equation

$$t^4 - a_1 t^3 + a_2 t^2 - a_3 t + a_4 = 0,$$

then we have seen (§1, D) that $s_1 = x_1 x_2 + x_3 x_4$, $s_2 = x_1 x_3 + x_2 x_4$, and $s_3 = x_1 x_4 + x_2 x_3$ are the roots of a cubic equation

$$t^3 - b_1 t^2 + b_2 t - b_3 = 0,$$

where $b_1 = a_2$, $b_2 = a_1 a_3 - 4 a_4$, and b_3 is a polynomial in a_1, a_2, a_3, a_4 which we do not write down. To calculate x_1 and x_2, we will calculate $x_1 + x_2$ and $x_1 x_2$.

In the first place, $(x_1 x_2) + (x_3 x_4) = s_1$ and $x_1 x_2 x_3 x_4 = a_4$, and so $x_1 x_2$ is a root of the equation

(1) $$t^2 - s_1 t + a_4 = 0.$$

In the second place, let us consider the polynomial

$$y_1 = x_1 + x_2 - x_3 - x_4.$$

Since $x_1^2 + x_2^2 + x_3^2 + x_4^2 = a_1^2 - 2a_2$, we have

$$y_1^2 = a_1^2 - 2a_2 + 2(s_1 - s_2 - s_3) = a_1^2 - 4a_2 + 4s_1,$$

and so we obtain y_1 by taking a square root.

Finally, since $x_1 + x_2 + x_3 + x_4 = a_1$, we have

$$x_1 + x_2 = \frac{1}{2}(a_1 + y_1).$$

If α is a root of (1), then x_1 and x_2 are the roots of

$$t^2 - \frac{1}{2}(a_1 + y_1)t + \alpha = 0.$$

Then we have $x_3 x_4 = s_1 - \alpha$ and $x_3 + x_4 = (a_1 - y_1)/2$, and so x_3 and x_4 are the roots of

$$t^2 - \frac{1}{2}(a_1 - y_1)t + s_1 - \alpha = 0.$$

2. Additional Remarks on Groups and on the Resolution of Algebraic Equations

A. The Symmetric Group $\mathfrak{S}_n$

Let E and F be two finite sets, each having n elements. By induction on n, the number of *bijections* $f : E \to F$ (§4, A) is equal to $n!$. Indeed, let a be an element of E and b an arbitrary element of F. For each bijection g from $E \setminus \{a\}$ onto $F \setminus \{b\}$, there corresponds to g exactly one bijection f from E onto F with $f(x) = g(x)$ for $x \in E \setminus \{a\}$ and with $f(a) = b$, and this gives $(n-1)!$ bijections because of the inductive hypothesis. Since there are n ways of choosing b in F, the number of bijections from E into F is $n \cdot (n-1)! = n!$ by definition. If we take $E = F = \{1, \ldots, n\}$, the bijections from E onto itself are the Cauchy "substitutions", and so their number is $n!$.

B. The Galois Group of an Equation

The originality of Galois compared to his predecessors was in his associating with *every* algebraic equation $P(t) = 0$ without multiple roots and whose coefficients belong to an arbitrary *subfield* K of $\mathbb{C}$ (§3, C III), a *group* of permutations, which is in general *different* from the symmetric group. Let $z_1, z_2, \ldots, z_n$ be the distinct roots of $P(t) = 0$, so that

$$P(t) = a_0(t - z_1)(t - z_2) \ldots (t - z_n).$$

The subfield $E = K(z_1, z_2, \ldots, z_n)$ of $\mathbb{C}$ *generated* by $z_1, z_2, \ldots, z_n$ *(loc.cit.)* is called the *root field* of P. We recall that the elements of E are the values

$$(1) \qquad \frac{R(z_1, z_2, \ldots, z_n)}{S(z_1, z_2, \ldots, z_n)}$$

of rational fractions with coefficients in K for which $S(z_1, z_2, \ldots, z_n) \neq 0$. For *each* permutation $\sigma \in \mathfrak{S}_n$, the number

$$(2) \qquad \frac{R(\sigma(z_1), \sigma(z_2), \ldots, \sigma(z_n))}{S(\sigma(z_1), \sigma(z_2), \ldots, \sigma(z_n))}$$

thus belongs to E provided that $S(\sigma(z_1), \sigma(z_2), \ldots, \sigma(z_n)) \neq 0$. But it can happen that, for a polynomial R, we have

$$R(z_1, z_2, \ldots, z_n) = 0 \quad \text{but} \quad R(\sigma(z_1), \sigma(z_2), \ldots, \sigma(z_n)) \neq 0.$$

The set G of permutations $\sigma \in \mathfrak{S}_n$ such that, for *every* polynomial R with $R(z_1, z_2, \ldots, z_n) = 0$, *one also has* $R(\sigma(z_1), \sigma(z_2), \ldots, \sigma(z_n)) = 0$ is called the *Galois group* of the equation $P(t) = 0$ (or of the polynomial P, or of the field E). It is clear that G is indeed a *subgroup* of $\mathfrak{S}_n$, for if σ and τ belong to G, then the same is true of their composition $\sigma\tau$, by the definition of G.

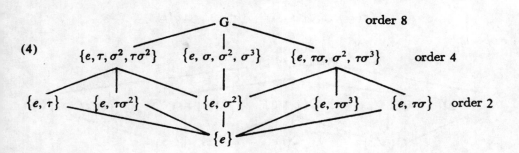

(4)

Example

Take $K = \mathbf{Q}$, and consider the equation

(3)
$$t^4 - 2 = 0.$$

Since $i^4 = (-1)^2 = 1$, the four roots are

$$z_1 = \sqrt[4]{2}, \qquad z_2 = i\sqrt[4]{2}, \qquad z_3 = -\sqrt[4]{2}, \qquad z_4 = -i\sqrt[4]{2},$$

and there are numerous polynomial relations with rational coefficients between these roots, for example

$$z_1 z_2 - z_3 z_4 = 0.$$

It is thus easy to give examples of permutations in $\mathfrak{S}_4$ which do not belong to the Galois group G, for example the cyclic permutation $(z_2 z_3 z_4)$ (§2, B), because $z_1 z_3 - z_4 z_2 \neq 0$.

It can be shown that G is *generated* by the two cyclic permutations

$$\sigma = (z_1 z_2 z_3 z_4) \qquad \text{and} \qquad \tau = (z_2 z_4),$$

which are such that

$$\sigma^4 = e, \qquad \tau^2 = e, \qquad \sigma\tau = \tau\sigma^3.$$

This shows that the group G is not commutative. The group has eight elements, namely

$$e, \quad \sigma, \quad \sigma^2, \quad \sigma^3, \quad \tau, \quad \tau\sigma, \quad \tau\sigma^2, \quad \tau\sigma^3,$$

and the collection of subgroups of G, ordered by inclusion, is easily described by diagram (4).

C. Galois Groups and Groups of Automorphisms

The Galois group G of a root field E *operates* on E (§3, F). If x is an element (1) of E, then $\sigma \cdot x$ is the element (2). It is of course necessary to verify that, if an element of E is written in two ways $R(z_1, \ldots, z_n)$ and $S(z_1, \ldots, z_n)$ for different rational fractions R and S, then we still have

$$R(\sigma(z_1),\dots,\sigma(z_n)) = S(\sigma(z_1),\dots,\sigma(z_n)).$$

(5)

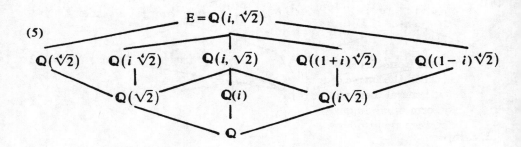

But this is equivalent to saying, on setting $H = R - S$, that the relation $H(z_1,\dots,z_n) = 0$ entails that $H(\sigma(z_1),\dots,\sigma(z_n)) = 0$, and this is the definition of the group G. It is immediate that $e \cdot x = x$ and $(\sigma\tau) \cdot x = \sigma \cdot (\tau \cdot x)$. In particular, $\sigma^{-1} \cdot (\sigma \cdot x) = e \cdot x = x$, which shows that $x \mapsto \sigma \cdot x$ is a *bijection* from E onto itself. Further, we have $\sigma \cdot (x+y) = \sigma \cdot x + \sigma \cdot y$ and $\sigma \cdot (xy) = (\sigma \cdot x)(\sigma \cdot y)$; in other words, $x \mapsto \sigma \cdot x$ is an *isomorphism* from the field E onto itself. Such a map is called an *automorphism* of E. The fundamental theorem of Galois theory is that there is a *bijection* from the set of *subgroups* of G onto the set of *subfields* of E such that $K \subset L \subset E$. Corresponding to a subgroup H of G is the subfield L of elements of E *invariant* under the action of H, and the subgroup H of G formed by the permutations which leave each element of L invariant corresponds to a subfield L. If $H' \subset H$, then the subfield L' which corresponds to H' contains L. For example, for the root field E of equation (3), the following diagram of *subfields* of E corresponds to the diagram 4 of subgroups of the Galois group.

D. Normal Subgroups and Simple Groups

Jordan was apparently the first mathematician to use the word "isomorphism". He used the term *holohedral isomorphism* for what we now call an "isomorphism" from a group G onto a group G'. But Jordan also introduced another notion which became fundamental. He used the term *merihedral isomorphism* for a map $f : G \to G'$ such that:

1) $f(xy) = f(x)f(y)$ for $x \in G$ and $y \in G$;

2) each $x' \in G'$ is of the form $f(x)$ for at least one $x \in G$.

In case 2), we now say that f is a *surjection* and we write

$$f(G) = G'.$$

The maps f which satisfy only the *first* condition above are now called *homomorphisms* from G into G'. We then have $f(e) = e'$, the identity element of G', because

$$f(x) = f(ex) = f(e)f(x).$$

It follows that

$$f(x^{-1}) = f(x)^{-1}$$

since

$$e' = f(e) = f(xx^{-1}) = f(x)f(x^{-1}).$$

We conclude from this that, for each homomorphism f, $f(G)$ is a *subgroup* of G'. It is called the *image* of G under f.

But in general there will be elements $x \in G$ other than e such that $f(x) = e'$. They form a *subgroup* N of G, called the *kernel* of f. In fact, if $f(x) = f(y) = e'$, then we also have

$$f(xy) = f(x)f(y) = e' \quad \text{and} \quad f(x^{-1}) = f(x)^{-1} = e'.$$

But this subgroup N is not an arbitrary one. It possesses an additional property: for *every* $z \in G$ and $x \in N$ we again have

$$zxz^{-1} \in N$$

because $f(zxz^{-1}) = f(z)f(x)f(z^{-1}) = f(z)f(z^{-1}) = e'$.

The subgroups of G which have this property are called *normal* subgroups (we also say *distinguished* subgroups). They were introduced for the first time by Galois. It was he who, in the case of an arbitrary subgroup H of G, thought of arranging the elements of G in *left equivalence classes* "modulo H", as Cauchy had done (§2, B). These classes are sets of the form sH, formed of all the elements sh in which h runs through all the elements of H. If two classes sH and $s'H$ have a common element $sh = s'h'$, then we have $s' = shh'^{-1}$, and so for each $h'' \in H$,

$$s'h'' = shh'^{-1}h'' \in sH.$$

In other words, $s'H \subset sH$, and similarly $sH \subset s'H$, so that the two classes are identical. Thus G is the union of *distinct* classes such that no two classes have a common element. In the same way, Galois considered the right equivalence classes Hs. But in general $Hs \neq sH$. One has $Hs = sH$ for each $s \in G$ only if $sHs^{-1} = H$, which is to say that the subgroup H is *normal*.

When N is a normal subgroup, we can define a law of composition between equivalence classes, for we have $(sN)(tN) = s(Nt)N = (st)N$ because of the equalities $tN = Nt$ and $NN = N$. It is easily verified that this law defines a group structure on the set (written G/N) of equivalence classes. This group is called the *quotient group* of G by N. It is also easily verified that the map $f : s \mapsto sN$ is a surjective homomorphism from G onto G/N, whose kernel is N. When G is finite, the order of G/N is the quotient of the order of G by the order of G/N, and is denoted by $(G : N)$.

If a group G is commutative, then all the subgroups of G are normal. For example, if G is the additive group $\mathbf{Z}$ of integers and N the subgroup $m\mathbf{Z}$ of multiples of m, where $m > 1$, then $\mathbf{Z}/m\mathbf{Z}$ is the additive group of congruence classes modulo m (§2, C).

We say that a group G is *simple* if it does not contain any normal subgroup other than G itself and the subgroup $\{e\}$ containing just the identity element. The only finite *commutative* groups which are simple are the *cyclic groups of prime order*. In fact, let G be a commutative group of order $n = rs$, where $r > 1$ and $s > 1$. For each $x \neq e$ in G, we have $x^n = e$, which is written $(x^r)^s = e$. If $x^r = e$, the subgroup generated by x is not simply $\{e\}$, and it has an order which divides r, and so it is different from G. If $y = x^r \neq e$, we have $y^s = e$, and the same conclusion follows. A finite and commutative simple group G is thus of prime order p. Further, if $x \neq e$, all the powers x^h for $1 \leq h \leq p - 1$ are distinct, for otherwise we would have $x^{h-k} = e$ for $1 < h - k < p$, and $h - k$ would have to divide p, which is absurd.

Jordan and Hölder showed that for *every* finite group G, there exists a sequence of distinct subgroups

$$G = G_0 \supset G_1 \supset G_2 \supset \ldots \supset G_{r-1} \supset G_r = \{e\}$$

such that each G_{k+1} is *normal* in G_k and such that each of the groups G_k/G_{k+1} is *simple*.

In the example (4), the three subgroups of order 4 are normal in G and are commutative, and so each chain of subgroups joining G and $\{e\}$ is a sequence (6) of Jordan-Hölder. When all the simple groups G_k/G_{k+1} are commutative (and hence cyclic of prime order), it is said that G is a *solvable* group. It can be proved that groups whose order is divisible only by one or two prime numbers and groups whose order is not divisible by the square of a prime are solvable. The most remarkable theorem of this type is the theorem of Feit and Thompson: each finite group of *odd* order is solvable; the proof of this theorem proceeds by contradiction and consists of more than 250 pages! It was Galois who introduced the notion of solvable group, and this gave him the key to the problem of the resolution of an algebraic equation "by radicals": the necessary and sufficient condition for such a resolution to exist is that the Galois group of the equation be solvable.

E. Rotations of the Cube

The definition of a rotation around a line Δ which is not the identity rotation shows the following two facts: if a point M is *invariant* with respect to the rotation, it is on Δ; and if the transform M' of M is distinct from M, then the plane perpendicular to MM' through its midpoint contains Δ.

With the aid of these two remarks, it is easy to determine all the rotations which leave a cube $K = ABCDA'B'C'D'$ invariant. The centre O of the cube is invariant, and so the axis of such a rotation always passes through O. If a vertex A is not invariant, it is transformed either to an extreme point of an edge issuing from A, or to the vertex opposite to A on one of the faces passing through A, or finally to the vertex A' which is symmetric to A with respect to O. We examine the four possible cases using the fact that the line OE, where E is the midpoint of AB (Figure 40), is perpendicular to the plane $CDC'D'$.

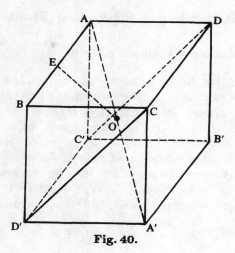

Fig. 40.

It is easy to see that the rotations which we are seeking are the following, together with the identity rotation:

1) The rotations of 180° around a line joining O to the middle of an edge; there are 6 of these. Each of these forms, together with the identity mapping, a cyclic subgroup of order 2.

2) The rotations of 120° and 240° around a diagonal of the cube; for each diagonal, these together with the identity mapping form a cyclic subgroup of order 3. There are 8 such rotations.

3) The rotations of 90°, 180° and 270° around a line perpendicular to a face; for each of these lines, they form together with the identity mapping a cyclic subgroup of order 4. There are 9 such rotations.

The total number of rotations is thus $1 + 6 + 8 + 9 = 24$.

The group G of rotations of the cube *operates* on the three perpendiculars leading from O to the faces of the cube, and it is easily seen that G permutes them in every possible way. If each rotation $\sigma \in G$ is associated with the corresponding permutation of these three lines, a surjective homomorphism from G onto the group $\mathfrak{S}_3$ of permutations of these lines is defined. The kernel N of this homomorphism is thus a normal subgroup of order 4 of G formed by the identity and the rotations of 180° around the three lines. A second normal subgroup $T \supset N$ of order 12 is generated by N and the rotations around the four diagonals of the cube; it is the subgroup which leaves the three lines invariant or which permutes them without leaving any of them fixed[1].

---

[1] The points A, C, B', D' are the vertices of a regular tetrahedron, and the subgroup T leaves this tetrahedron invariant. The other 12 rotations of G transform the tetrahedron $ACB'D'$ into its symmetric image $A'C'BD$ with respect to O.

3. Additional Remarks on Rings and Fields

A. Congruences Modulo a Prime Number

We have seen (§2, C) that, for each integer $m > 1$, the *addition* $(\overline{r}, \overline{s}) \mapsto \overline{r} + \overline{s}$ of congruence classes modulo m defines a structure of a *cyclic group* of order m on the set of these classes. To say that (by definition) $\overline{r} + \overline{s} = \overline{r + s}$ means that the map $r \mapsto \overline{r}$ is a surjective homomorphism (Appendix 2, D) of the additive group $\mathbf{Z}$ of integers onto the group of congruence classes. The kernel of this homomorphism is the subgroup $m\mathbf{Z}$ of multiples of m, and the additive group of congruence classes modulo m can thus be denoted by $\mathbf{Z}/m\mathbf{Z}$. But there is a second law of composition on $\mathbf{Z}/m\mathbf{Z}$, namely, $(\overline{r}, \overline{s}) \mapsto \overline{r} \cdot \overline{s} = \overline{rs}$ (§1, F); which is associative and commutative, and it is immediately verified that we have

$$\overline{r} \cdot (\overline{s} + \overline{t}) = \overline{r} \cdot \overline{s} + \overline{r} \cdot \overline{t}.$$

This second law thus defines on $\mathbf{Z}/m\mathbf{Z}$ a structure of a commutative *ring* which has an identity element $\overline{1}$. Let us suppose now that we take the ring $\mathbf{Z}/p\mathbf{Z}$, where p is a *prime* number. Then the ring $\mathbf{Z}/p\mathbf{Z}$ is a commutative *field* denoted by $\mathbf{F}_p$. In fact, if $0 < r < p$, then the greatest common divisor of p and r is 1, and so we have Bezout's identity (see Chapter IV, Appendix 3)

$$mr + np = 1$$

for two integers m and n. Since $\overline{p} = \overline{0}$ in $\mathbf{Z}/p\mathbf{Z}$, it follows from this that $\overline{m} \cdot \overline{r} = \overline{1}$. In other words, each class $\overline{r} \neq \overline{0}$ in $\mathbf{Z}/p\mathbf{Z}$ has an *inverse*.

The set $\mathbf{F}_p^*$ of classes $\overline{r} \neq \overline{0}$ in $\mathbf{F}_p$ is thus (as in all fields) a *multiplicative group*, of order $p - 1$. For each $\overline{r} \neq \overline{0}$, we thus have $\overline{r}^{p-1} = \overline{1}$. In terms of congruences, this is written

$$r^{p-1} \equiv 1 \pmod{p}$$

for each natural number r which is not a multiple of p. This is one of the first theorems of number theory, proved by Fermat and known under the name of Fermat's "little" theorem.

But one can say much more: the group $\mathbf{F}_p^*$ is *cyclic*, a result conjectured by Euler and proved for the first time by Gauss (using a different terminology). It can be supposed that $p \neq 2$ and three lemmas are used:

I) Let n be an arbitrary natural number with $n > 1$. For each natural number $k \leq n$, let d be the greatest common divisor of k and n. Then the *order* of the class $\overline{k}$ in the group $\mathbf{Z}/n\mathbf{Z}$ is equal to n/d. In fact, let $n = n'd$ and $k = k'd$. If $ak = ak'd$ is divisible by $n = n'd$, then n' divides ak', and, since the greatest common divisor of n' and k' is 1, n' divides a. In particular, the only natural numbers $k \leq n$ such that $\overline{k}$ *generates* $\mathbf{Z}/n\mathbf{Z}$ are the natural numbers which are "coprime to n", that is to say the natural numbers whose greatest common divisor with n is 1. The number of these is designated by $\varphi(n)$, which is called *Euler's function* of n.

II) The relation "d divides n" is written $d \mid n$. For each $n > 1$, we have

$$\sum_{d \mid n} \varphi(d) = n. \tag{1}$$

In fact, for each divisor d of n, if $n = n'd$, the class $\overline{n'}$ is of order d in the group $\mathbf{Z}/n\mathbf{Z}$, and from I) there are exactly $\varphi(d)$ natural numbers $k \leq n$ such that $\bar{k}$ is of order d. Since for $1 \leq k \leq n$, the order of $\bar{k}$ is a divisor of n, the relation (1) follows. In addition it is seen that for each divisor d of n, $\mathbf{Z}/n\mathbf{Z}$ has *only one* subgroup of order d.

Conversely, if a finite group G of order n is such that, for each divisor d of n, there are *at most* d elements x such that $x^d = e$, then G is cyclic. In fact, if there exists an element y of order d, then all the elements of the subgroup $\{e, y, y^2, \ldots, y^{d-1}\}$ are such that $x^d = e$, and from I) there are exactly $\varphi(d)$ elements of G which are of order d. If there existed a divisor d of n such that *no* element of G were of order d, then n would be the sum $\sum \varphi(d')$, where d' runs through a set of divisors *different* from d. But this would contradict (1). In particular, the existence of at least one element of order n implies that G is cyclic.

III) Let K be an arbitrary commutative field, and let

$$P(t) = t^q + a_1 t^{q-1} + \ldots + a_q$$

be a polynomial of degree q with coefficients in K. Then there are *at most* q roots of $P(t) = 0$ in K. This can be seen by induction on q: let us suppose that $P(c) = 0$ for some $c \in K$, and consider the polynomial

$$P(t + c) = (t + c)^q + a_1(t + c)^{q-1} + \ldots + a_q,$$

which is written

$$P(c) + tQ(t)$$

for some polynomial $Q(t)$ of degree $q - 1$. Since $P(c) = 0$, it can be seen that we can have $P(c') = 0$ for $c' \neq c$ only if $Q(c' - c) = 0$, which gives at most $q - 1$ values of $c' \in K$.

If we return to the finite field $\mathbf{F}_p$, from III), for each divisor d of $p - 1$ there are at most d elements of $\mathbf{F}_p^*$ such that $x^d = 1$. As a consequence of II), the multiplicative group $\mathbf{F}_p^*$ is cyclic. Its generators $\bar{g}$, of which the number is $\varphi(p - 1)$, were called "primitive roots" by Gauss. Each element of $\mathbf{F}_p^*$ can therefore be written in a unique manner $\bar{g}^k$ for a natural number k with $1 \leq k \leq p - 1$. The elements of $\mathbf{F}_p^*$ which are *squares* are the elements $\bar{g}^{2k}$ for $1 \leq k \leq (p - 1)/2$; the only element $\overline{-1}$ different from the identity element whose square is the identity element is $\bar{g}^{(p-1)/2}$. From this we conclude that $\overline{-1}$ is a square if and only if $(p - 1)/2$ is *even*, which is to say that the prime number p is of the form $4k + 1$. In terms of congruences, the congruence

$$z^2 \equiv -1 \pmod{p}$$

has a solution only if p is of the form $4k + 1$.

B. The Ring $Z[i]$ of Gaussian Integers

Let us recall first that for each complex number $z = x + iy$, the complex number $x - iy$ is called the *conjugate* of z and is denoted by $\bar{z}$. We have $z + \bar{z} = 2x$ and $z - \bar{z} = 2iy$, so that $z = \bar{z}$ if and only if z is real. The number $z\bar{z} = x^2 + y^2$ is called the *norm* of z and is denoted by $N(z)$. It is the square $|z|^2$ of the absolute value. We have $\overline{zz'} = \bar{z} \cdot \overline{z'}$, and

$$(2) \qquad\qquad N(zz') = N(z)N(z').$$

The number $N(z - z') = |z - z'|^2$ is thus the square of the *distance* between the points z and z' in the plane $\mathbf{R}^2$.

We have seen (§3, C III) that the numbers $x + iy$ where x and y are *rational* form a *field* $\mathbf{Q}(i)$, and those among these numbers for which x and y are *integers* form a *ring* $Z[i]$ whose elements are called *Gaussian integers*. To avoid confusion, the ordinary integers (positive and negative) are called *rational integers*. For each Gaussian integer z, $N(z)$ is a non-negative integer. The Gaussian integers are the "nodes" of the *lattice* of the plane formed by the parallels to the two axes passing through the points of the axes with integral coordinates (Figure 41).

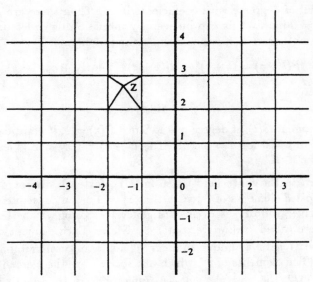

Fig. 41.

The property on which all the "arithmetic" of the Gaussian integers rests is the following: *for each number $z \in \mathbf{Q}(i)$, there exists at least one Gaussian integer u such that*

(3)
$$N(z - u) \leq \frac{1}{2}.$$

This is immediate because z is in one of the squares (including the sides) of the lattice, and there is always one of the four vertices whose distance to z is at most the semi-diagonal $1/\sqrt{2}$ (Figure 41).

It follows from this that in the ring $\mathbf{Z}[i]$ there is a process of *Euclidean division* : if u and v are two Gaussian integers such that $v \neq 0$, then there are two Gaussian integers q and r such that

(4)
$$u = qv + r \quad \text{with} \quad N(r) \leq \frac{1}{2}N(v).$$

In fact, it suffices to apply (3) to the number u/v in $\mathbf{Q}(i)$. This shows that there is a Gaussian integer q such that

$$N\left(\frac{u}{v} - q\right) \leq \frac{1}{2},$$

and from this it is deduced, using (2), that

$$N(u - qv) = N(v)N\left(\frac{u}{v} - q\right) \leq \frac{1}{2}N(v).$$

All the arguments of Euclidean arithmetic rest on Euclidean division and the fact that, if there is a decreasing sequence

$$r_1 > r_2 > \ldots > r_n > \ldots$$

of natural numbers which can only stop if a term is 0, then this sequence has only finitely many terms and the last one is 0. It is sufficient to apply this remark to a sequence of *norms* of Gaussian integers

$$N(r_1) \geq 2N(r_2) \geq \ldots \geq 2^{n-1}N(r_n) \geq \ldots$$

to see that the sequence is finite and terminates with $r_n = 0$.

The properties of the rational integers can thus be replicated essentially without change for the Gaussian integers. It is simply necessary to take account of the fact that, in the ring $\mathbf{Z}[i]$, there are not only the numbers 1 and -1, but also the numbers i and $-i$, which have an inverse in $\mathbf{Z}[i]$. Moreover, these are the only such numbers, for, if z has an inverse z', then we have $N(z)N(z') = 1$, whence $N(z) = 1$ because $N(z)$ is a non-negative integer. We say that the four integers ± 1 and $\pm i$ are the *units* of $\mathbf{Z}[i]$.

The *prime* Gaussian integers are, by definition, the integers u other than the four units such that, if $u = vw$ is the product of two Gaussian integers, one of them is a unit. For each prime Gaussian integer u, the four Gaussian integers $\pm u$ and $\pm iu$ are prime; we say that they are *associates*. Among them is just one $u = a + ib$ with $a > 0$ and $b \geq 0$; these are said to be the prime Gaussian integers *of the first quadrant*. If $u \in \mathbf{Z}[i]$ is prime, the same is true for $\overline{u}$, but the only non-real, prime integer of the first quadrant which is associated with its conjugate $\overline{u}$ is $u = 1 + i$, because $1 - i = -i(1 + i)$.

The fundamental result is *the theorem of decomposition into prime integers*: every Gaussian integer z can be written *in a unique way*, as

$$(5) \qquad z = \varepsilon u_1^{\alpha_1} u_2^{\alpha_2} \ldots u_k^{\alpha_k},$$

where the u_j are distinct prime integers of the first quadrant, the α_j are natural numbers, and ε is a unit.

Two corollaries follow from this:

a) If a prime integer divides a product of two integers, then it divides one of them.

b) If two non-associated prime integers each divide the same integer, then their product divides this integer.

If two integers z and z' are specified by their decompositions into prime integers, (5) and

$$(6) \qquad z' = \varepsilon' u_1^{\beta_1} u_2^{\beta_2} \ldots u_h^{\beta_h},$$

then their greatest common divisor is obtained by taking the prime integers $w_1, \ldots, w_l$ which occur in *both* of the decompositions (5) and (6) and by forming the product

$$d = w_1^{\gamma_1} w_2^{\gamma_2} \ldots w_l^{\gamma_l},$$

where each exponent γ_j is the smaller of the two exponents of w_j which occur in (5) and (6). We have an identity of Bezout

$$(7) \qquad d = az + bz',$$

where a and b are two integers in $\mathbf{Z}[i]$: this has the same proof, based on Euclidean division, as Bezout's identity in $\mathbf{Z}$ (see Chapter IV, Appendix 3).

The relations between the prime numbers of $\mathbf{Z}$ and those of $\mathbf{Z}[i]$ can be described comprehensively.

I) The prime number 2 is (to within a unit) the *square* of a prime integer because

$$2 = (1+i)(1-i) = -i(1+i)^2.$$

II) If a rational prime $p \neq 2$ is not a prime in $\mathbf{Z}[i]$, it is divisible by a non-real prime integer u of the first quadrant, and hence also by its conjugate $\bar{u}$. Since $p \neq 2$, u and $\bar{u}$ are not associates, and consequently p is divisible by $u\bar{u} = N(u)$. But since p is prime in $\mathbf{Z}$, it necessarily follows that $p = N(u)$. If we set $u = x + iy$, where $x > 0$ and $y > 0$, we thus have $p = x^2 + y^2$, which implies that p has the form $4k + 1$.

III) If w is prime in $\mathbf{Z}[i]$, we define congruences modulo w in $\mathbf{Z}[i]$ as in A), and it follows from Bezout's identity (7) that each class $\neq \bar{0}$ in the ring $\mathbf{Z}[i]/w\mathbf{Z}[i]$ has an inverse. In other words, $\mathbf{Z}[i]/w\mathbf{Z}[i]$ is a *field*. If p is a rational prime of the form $4k + 1$, we have seen in A) that there is a rational integer x such that $x^2 + 1 \equiv 0 \pmod{p}$. But this means that in $\mathbf{Z}[i]$ we have

$$(x + i)(x - i) \equiv 0 \quad (\text{mod } p),$$

and since we also see that neither $x + i$ nor $x - i$ is divisible by p in $Z[i]$, the equivalence classes of $x + i$ and $x - i$ in $Z[i]/pZ[i]$ are divisors of zero. Hence $Z[i]/pZ[i]$ is not a field, and so p *is not a prime* in $Z[i]$. Now take u to be a prime integer in the first quadrant which divides p. It cannot belong to Z, and so it follows from II) that $p = N(u)$, and u is the only prime integer of the first quadrant which divides p. If we set $u = a + ib$, then we have $p = a^2 + b^2$, and we thus see that every prime number of the form $4k + 1$ is in a unique way the sum of the the the squares of two natural numbers: this is the theorem of Fermat-Euler (see Chapter IV, Appendix 3).

When Euclidean division can be generalized in the same way in the ring $Z\left[\sqrt{D}\right]$, totally analogous theorems can be deduced. Unfortunately, it has been shown that this is only possible for *finitely many* integers D not divisible by a square. For $D < 0$, the possible values of D are -1, -3, -7, and -11. In fact, there are infinitely many values of D for which *there is not a unique decomposition into prime factors*. For example, for $D = -5$, we have

$$9 = 3 \cdot 3 = \left(2 + \sqrt{-5}\right)\left(2 - \sqrt{-5}\right),$$

and it can be proved that 3, $2 + \sqrt{-5}$, and $2 - \sqrt{-5}$ are prime. The theory of algebraic numbers was born with the discovery by Kummer, Kronecker and Dedekind that it is possible to re-establish a theorem of "unique decomposition" by introducing new objects, *ideals* and *valuations*.

C. Congruences modulo a Polynomial

Here we shall give a third example of the "calculus of congruences". Let K be a field with infinitely many elements. The *polynomials* with coefficients in K are the maps

$$(8) \qquad\qquad x \mapsto P(x) = a_0 x^n + \ldots + a_n,$$

which are the sum of a finite number of *monomials* $a_j x^{n-j}$. The polynomial P is said to be of *degree* n (we write $\deg(P) = n$) if the monomial $a_0 x^n$ of highest degree in P has a non-zero coefficient. The polynomials of degree 0 are the non-zero *constants* $c \in K$. It is immediate that the polynomials with coefficients in K form a commutative *ring*, denoted by $K[x]$ with respect to the sum and product of polynomials. Since K is infinite, a polynomial (8) can be identically zero only if $a_0 = a_1 = \ldots = a_n = 0$, because we have seen in A) that a polynomial of degree n cannot be zero at more than n elements of K. Let us note that the product

$$(a_0 x^m + \ldots + a_m)(b_0 x^n + \ldots + b_n)$$

of a polynomial of degree m and of a polynomial of degree n is of degree $m + n$, because there is only one monomial of this degree in the product, namely $a_0 b_0 x^{m+n}$, and we have $a_0 b_0 \neq 0$ if $a_0 \neq 0$ and $b_0 \neq 0$ because K is a field.

We deduce from this that the only elements of $K[x]$ which have an *inverse* in $K[x]$ are the *constants* $c \neq 0$.

The fundamental property of this ring is that a process of *Euclidean division* can again be defined on it: if A and B are two polynomials such that $B \neq 0$, then there exist two polynomials Q and R such that we have

(9) $A = BQ + R$,   with $R = 0$ or $\deg(R) < \deg(B)$.

In fact, let m and n be the degrees of A and B, respectively. If $A = 0$ or if $m < n$, then it is sufficient to take $Q = 0$ and $R = A$ to obtain (9). If this is not the case, a monomial $M_1 = c x^{m-n}$ can be found such that, if

(10) $A_1 = A - B M_1$,   then we have $A_1 = 0$ or $\deg(A_1) < \deg(A)$.

In fact, if $A = a_0 x^m + \ldots + a_m$ and $B = b_0 x^n + \ldots + b_n$, it is sufficient to take $c = a_0 / b_0$ to obtain (10).

The procedure is repeated, and we thus have a sequence of polynomials:

$$A_1 = A - B M_1 \quad \text{with} \quad A_1 = 0 \quad \text{or} \quad \deg(A_1) < \deg(A),$$
$$A_2 = A_1 - B M_1 \quad \text{with} \quad A_2 = 0 \quad \text{or} \quad \deg(A_2) < \deg(A_1),$$

etc., and we necessarily arrive at

$$A_{k+1} = A_k - B M_{k+1} \quad \text{with} \quad A_{k+1} = 0 \quad \text{or} \quad \deg(A_{k+1}) < n$$

after a finite number of steps at most equal to the difference $m - n + 1$. We thus have

$$A_{k+1} = A - B(M_1 + M_2 + \ldots M_{k+1}),$$

which gives the equation (9) with $Q = M_1 + M_2 + \ldots + M_{k+1}$ and $R = A_{k+1}$.

The polynomials Q and R satisfying (9) are uniquely determined. In fact, if it were also the case that $A = BQ_1 + R_1$ with $R_1 = 0$ or $\deg(R_1) < \deg(B)$, then it would follow that

$$B(Q - Q_1) = R_1 - R,$$

and, since $\deg(R_1 - R) < \deg(B)$ if $R_1 - R \neq 0$, this results in a contradiction in the case where $Q - Q_1 \neq 0$. Thus necessarily $Q - Q_1 = 0$ and $R - R_1 = 0$. It is said that Q is the *quotient* and that R is the *remainder* of the *Euclidean division* of A by B.

All of Euclid's arithmetical arguments are thus transcribed forthwith. That which corresponds to the prime numbers are here the *irreducible* polynomials, that is to say, the non-constant polynomials P such that, if $P = AB$ is the product of two polynomials, then one of them is necessarily a constant. If P is irreducible, then the same is true for cP for each non-zero $c \in K$. Among these polynomials we choose the one which is *monic*, that is to say, the one whose coefficient of the monomial of highest degree is 1. The theorem of unique decomposition into prime factors is thus that *each* non-constant polynomial A in $K[x]$ can be written uniquely in the form

$$(11) \qquad A = c P_1^{\alpha_1} P_2^{\alpha_2} \ldots P_k^{\alpha_k} ,$$

where $c \in K$, the P_j are distinct monic, irreducible polynomials, and the α_j are natural numbers. As in B), we can deduce the definition of a greatest common divisor (defined to within a constant factor) and Bezout's identity for two non-zero polynomials A and B, namely

$$(12) \qquad D = MA + NB ,$$

where D is a greatest common divisor of A and B, and M and N are two polynomials.

Next, a *congruence* $A \equiv B \mod M$ modulo a non-constant polynomial M is defined. A *class* $\bar{A}$ of polynomials congruent to a polynomial A modulo M is the set of polynomials $A + QM$, where Q is an arbitrary polynomial. These classes form a *ring* $K[x]/MK[x]$. The Euclidean division shows that each class $\bar{A}$ has a unique "representative" which is either 0 or the remainder of the Euclidean division of A by M, a polynomial of degree less than $\deg(M)$. Bezout's identity (12) also shows that, if P is an irreducible polynomial, then the ring $K[x]/PK[x]$ is a *field*.

D. The Field of Algebraic Functions

Given two polynomials $x \mapsto P(x)$ and $x \mapsto Q(x)$ over an infinite field K, with Q not identically zero, a "natural" definition of the "fraction" P/Q would be the function

$$x \longmapsto P(x)/Q(x) .$$

But this function is not defined at the points $x \in K$ such that $Q(x) = 0$. There are only finitely many of these points, but they depend on Q, and their presence is the cause of difficulties in the definition of operations on these "fractions". For example, the functions

$$x \longmapsto \frac{1}{x(x-1)} \quad \text{and} \quad x \longmapsto \frac{1}{x}$$

are not defined for $x = 0$, but since

$$\frac{1}{x(x-1)} + \frac{1}{x} = \frac{1}{x-1}$$

for $x \neq 1$, their sum is defined for $x = 0$. It is thus preferable not to speak of "values" for a "rational fraction" P/Q, but to consider that it is a *class of pairs* (A, B) of polynomials of $K[x]$ such that $B \neq 0$, two pairs (A, B) and (A_1, B_1) being in the *same* class if

$$AB_1 - BA_1 = 0 .$$

It is then necessary to define the sum and the product of two classes. For two pairs (A, B) and (C, D), we consider the pairs $(AD + BC, BD)$ and (AC, BD). It is necessary to show that, if we replace, for example, (A, B) by a pair (A_1, B_1) of the same class, the classes of the pairs "sum" and "product" do not change. This is a consequence of the identities

$$(AD + BC)B_1 D - (A_1 D + B_1 C)BD = (AB_1 - A_1 B)D^2 = 0$$

$$ACB_1 D - BDA_1 C = (AB_1 - A_1 B)CD = 0.$$

By abuse of language, we write A/B for the class of the pair (A, B), and we thus have $A/B = A_1/B_1$ for two pairs of the same class. The verification of the fact that these clases so defined form a *field* is immediate, the "inverse" of the class A/B being B/A in the case where $A \neq 0$. This field is denoted by $K(x)$, and it is said to be the *field of rational fractions* in one variable with coefficients in K.

If A and B have greatest common divisor D, so that $A = DA_1$ and $B = DB_1$, then we have $A/B = A_1/B_1$, and so a rational fraction can always be written as the quotient of two "coprime" polynomials (that is to say, such that their greatest common divisor is 1). But the sum of two rational fractions written in this form is not necessarily of this form.

Now suppose that K is the field $\mathbf{C}$ of complex numbers. An *algebraic plane curve* is the set Γ of points (x, y) of $\mathbf{C}^2$ satisfying an equation

(13) $$a_0(x)y^n + a_1(x)y^{n-1} + \ldots + a_n(x) = 0,$$

where the $a_j(x)$ are polynomials in $\mathbf{C}[x]$, with $a_0 \neq 0$. The curve Γ is said to be *irreducible* if the polynomial

$$P(y) = y^n + \frac{a_1(x)}{a_0(x)}y^{n-1} + \ldots + \frac{a_n(x)}{a_0(x)}$$

with coefficients in the field $\mathbf{C}(x)$ is irreducible. The ring $\mathbf{C}(x)[y]/P \cdot \mathbf{C}(x)[y]$ of congruence classes modulo P is then a *field L*, which is called the *field of rational functions on Γ* (or the field of *algebraic functions* on Γ). This name is explained in the following way: each element of L has a unique representation in the form

$$F = r_0(x)y^{n-1} + r_1(x)y^{n-2} + \ldots + r_{n-1}(x),$$

where the $r_j(x)$ are arbitrary elements of the field $\mathbf{C}(x)$ (these representatives are the *remainders* of the Euclidean divisions of polynomials of the ring $\mathbf{C}(x)[y]$ by P.)

Let (u, v) then be a point of $\mathbf{C}^2$ on the curve Γ, that is to say, satisfying equation (13). If u is not a zero of the denominators of the rational fractions $r_0, r_1, \ldots, r_{n-1}$, we can replace x by u and y by v in F. The complex number so obtained is the *value* of F at the point (u, v). It is possible to "extend" in a certain sense the definition of the "value" of F at the points where u is a zero of the denominators of the r_j and at what are called the "points at infinity" of

Γ in such a way as to obtain a true function. But this requires more advanced concepts relating to what are called "Riemann surfaces".

E. Remarks on Ordered Fields

Here we shall show that the field $\mathbf{R}(x)$ of rational fractions with real coefficients can be furnished with the structure of an *ordered field* (§3, F). We first define the relation $P \geq 0$ for a *polynomial*

$$(14) \qquad P(x) = a_0 x^n + a_1 x^{n-1} + \ldots + a_n .$$

Here $P \geq 0$ means by definition *either that* $P = 0$ (that is to say, $a_0 = a_1 = \ldots = a_n = 0$) *or that* $a_0 > 0$. If P and Q are two polynomials, the relation $P \leq Q$ means by definition that $Q - P \geq 0$ in the preceding sense. Finally, for a rational fraction $R = P/Q$, with $Q \neq 0$, we define the relation $R \geq 0$ to mean that $PQ \geq 0$ in the preceding sense, and for two rational fractions R and S, the relation $R \leq S$ means that $S - R \geq 0$.

We have to verify that the axioms of the structures of order (§3, D) and those of ordered fields (§3, F). We are quickly brought to three verifications.

1) The axiom of *total* order. It is sufficient to see that, if $R \geq 0$ and $-R \geq 0$, then $R = 0$, and in virtue of the preceding definitions, it suffices to consider the case of polynomials. But if the polynomial $P(x)$ of formula (14) is such that $P \geq 0$ and $-P \geq 0$, it is not possible that $P \neq 0$, for this would imply at the same time that $a_0 > 0$ and $a_0 < 0$ for a real number a_0, which is absurd.

2) If $R \geq 0$ and $S \geq 0$, then $RS \geq 0$. Here again the definitions bring the verification to the case of polynomials. But if $P \geq 0$ and $Q \geq 0$ for the polynomial (14) and the polynomial

$$(15) \qquad Q(x) = b_0 x^m + \ldots + b_m ,$$

then either we have one of the relations $P = 0$ and $Q = 0$, and hence $PQ = 0$, or we have $a_0 > 0$ and $b_0 > 0$, and the term of highest degree in PQ is $a_0 b_0 x^{m+n}$, with $a_0 b_0 > 0$. The same argument shows that $P^2 \geq 0$ for every polynomial P.

3) If $R \geq 0$ and $S \geq 0$, then $R + S \geq 0$. If $R = A/B$ and $S = C/D$, then we have $R + S = (AD + BC)/BD$, and it is a question of seeing that, if we have $AB \geq 0$ and $CD \geq 0$ for four polynomials A, B, C, D, then

$$(AD + BC)BD = ABD^2 + CDB^2 \geq 0 .$$

By 2), we are led to showing that, if two polynomials (14) and (15) are ≥ 0, then $P + Q \geq 0$. This is evident if $P = 0$ or $Q = 0$. If this is not the case, and if, for example, $m > n$, then the term of highest degree in $P + Q$ is $b_0 x^m$, with $b_0 > 0$. It only remains to consider the case where $m = n$: the term of highest degree in $P + Q$ is then $(a_0 + b_0)x^m$, and, since $a_0 > 0$ and $b_0 > 0$, we have $a_0 + b_0 > 0$.

The interest of this example is that Archimedes' Axiom (Chapter III, §6 and Appendix 2) *is not satisfied*. In fact, the real numbers are the polynomials of degree 0, and the relation $r \leq s$ for two real numbers is the same in $\mathbf{R}$ and in $\mathbf{R}(x)$. But the polynomial x is greater than r in $\mathbf{R}(x)$ for *every* real number r: it can be said that x is "infinitely large" with respect to the real numbers, and since then

$$0 < \frac{1}{x} < \frac{1}{r}$$

for each real number $r > 0$, the rational fraction $1/x$ is "infinitely small" in $\mathbf{R}(x)$.

We have just used the fact that, in $\mathbf{R}(x)$, the square of an arbitrary polynomial is ≥ 0. This is a general fact : for each ordered field K, and *every* element $x \in K$, we have $x^2 \geq 0$. Indeed, this follows directly from the axioms if $x \geq 0$; and, if $x < 0$, then we have $-x > 0$, and so $x^2 = (-x)^2 > 0$. This proves in particular that *there is no* structure of ordered field on the field $\mathbf{C}$ of complex numbers, because we have

$$i^2 = -1 = -(1)^2 \, ,$$

and at the same time we should have $i^2 > 0$ and $i^2 < 0$, which is absurd.

More generally, in an ordered field K, the relation

(16) $$x_1^2 + x_2^2 + \ldots + x_m^2 = 0$$

is only possible if $x_1 = x_2 = \ldots = x_m = 0$. Indeed, let us suppose that $x_1 \neq 0$, so that $x_1^2 > 0$. Since $x_2^2 + \ldots + x_m^2 \geq 0$, being the sum of positive elements, we have

$$0 < x_1^2 \leq x_1^2 + (x_2^2 + \ldots + x_m^2),$$

and thus (16) cannot occur. For example, *there is no* structure of an ordered field on the finite field $\mathbf{F}_p$ (Appendix 3, A) because in this field $\overline{1} = (\overline{1})^2$ and

$$\overline{0} = \overline{p} = \overline{1} + \overline{1} + \ldots + \overline{1} \quad (p \quad \text{terms}) \, .$$

Artin and Schreier have shown that the impossibility of a relation (16) for non-zero elements in a commutative field is not only necessary, but is also *sufficient* for the existence of the structure of an ordered field on K.

4. Examples of Distances

A. Distances on the Space of Continuous Functions

One of the most used "spaces of functions" in analysis is the space $C(0,1)$ of real-valued, continuous functions defined on the interval $0 \leq x \leq 1$. A distance (§3, E) denoted by $d_\infty(f,g)$ is defined on this space by the formula

(1) $$d_\infty(f,g) = \max\{|f(x) - g(x)| : 0 \leq x \leq 1\}$$

for two functions f and g of $C(0,1)$, where the right-hand side of (1) is the smallest real number $\alpha \geq 0$ such that $|f(x) - g(x)| \leq \alpha$ for **all** x such that $0 \leq x \leq 1$. If f, g, h are three functions of $C(0,1)$, we have, for all x,

$$f(x) - h(x) = (f(x) - g(x)) + (g(x) - h(x)),$$

and so

$$
\begin{aligned}
(2) \qquad |f(x) - h(x)| &\leq |f(x) - g(x)| + |g(x) - h(x)| \\
&\leq d_\infty(f,g) + d_\infty(g,h),
\end{aligned}
$$

whence, by definition,

$$(3) \qquad d_\infty(f,h) \leq d_\infty(f,g) + d_\infty(g,h).$$

This is Axiom (iii) for distances ("the triangle inequality") (*loc.cit.*); Axioms (i) and (ii) are obviously satisfied.

The concept of "neighbourhood" for this distance has a simple geometric interpretation : to say that $d_\infty(f,g) \leq \alpha$ means that the graph of g is contained in the "band" of breadth 2α formed of the points (x, y) such that $0 \leq x \leq 1$ and

$$f(x) - \alpha \leq y \leq f(x) + \alpha.$$

(See Figure 42.)

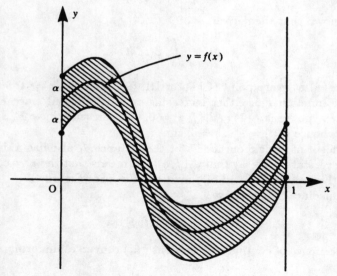

Fig. 42.

The concept of "limit" for this distance is very important in analysis. A sequence (f_n) of functions in $C(0,1)$ is said to *converge uniformly* to the function f when the sequence of numbers $(d_\infty(f, f_n))$ converges to 0.

These definitions are also valid for functions defined for $0 \leq x \leq 1$ and which are only supposed to be *bounded* (but not necessarily continuous). We then have a fundamental theorem: if a sequence (f_n) of *continuous* functions converges uniformly to a function f (necessarily bounded), then f is automatically *continuous*.

Another distance, denoted by $d_1(f,g)$, can be defined on $C(0,1)$ by the formula

(4)
$$d_1(f,g) = \int_0^1 |f(x) - g(x)|\, dx\,.$$

Here the triangle inequality

$$d_1(f,h) \leq d_1(f,g) + d_1(g,h)$$

also follows from the inequality (2), valid for all x in $[0,1]$, and from the fact that, if $u(x) \leq v(x)$ for two functions u and v of $C(0,1)$, then

$$\int_0^1 u(x)\, dx \leq \int_0^1 v(x)\, dx\,.$$

Only Axiom (i) for distances requires a proof : if $d_1(f,g) = 0$, then $f = g$. In fact, if

$$\phi(x) = |f(x) - g(x)|\,,$$

then it is known that the derivative of

$$F(x) = \int_0^x \phi(t)\, dt$$

is equal to $\phi(x)$ at every point (Chapter III, §9). Since $F'(x) \geq 0$ for each x in $[0,1]$, the function F is either increasing or constant. But since $F(0) = 0$ and since by hypothesis $F(1) = d_1(f,g) = 0$, we necessarily have $F(x) = 0$ for $0 \leq x \leq 1$, whence $\phi(x) = F'(x) = 0$, that is to say, $f = g$.

The concept of "neighbourhood" for the distance d_1 also has a simple geometric interpretation: to say that $d_1(f,g) \leq \alpha$ means that the *area* contained "between" the graphs of f and of g is at most α. (See Figure 43.)

It is immediate that we have

(5)
$$d_1(f,g) \leq d_\infty(f,g)\,.$$

Thus, if a sequence of continuous functions (f_n) converges uniformly to f, we also have

$$\lim_{n \to \infty} \int_0^1 f_n(x)\, dx = \int_0^1 f(x)\, dx$$

because

$$\left| \int_0^1 f(x)\, dx - \int_0^1 f_n(x)\, dx \right| = \left| \int_0^1 (f(x) - f_n(x))\, dx \right| \leq d_1(f, f_n)\,.$$

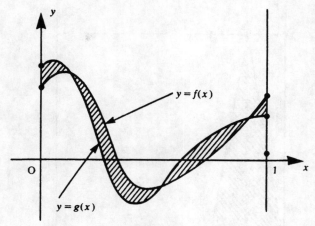

Fig. 43.

On the other hand, it can happen that the sequence (f_n) converges to f for the distance d_1, but *not* for d_∞ : it suffices to take

$$
\begin{cases}
f_n(x) = 2nx & (0 \le x \le \frac{1}{2n}), \\
f_n(x) = 2 - 2nx & (\frac{1}{2n} \le x \le \frac{1}{n}), \\
f_n(x) = 0 & (\frac{1}{n} \le x \le 1).
\end{cases}
$$

We have

$$
\int_0^1 f_n(x)\, dx = \frac{1}{2n},
$$

whence $(d_1(f_n, 0))$ converges to 0, although $d_\infty(f_n, 0) = 1$ for all n. (See Figure 44.)

B. Prehilbert Spaces

Let E be a vector space over the field **R** (§3, F). Generalizing the notion of scalar product in ordinary space (§1, B), let us suppose that we are given a map

$$(x, y) \mapsto (x \mid y)$$

from $E \times E$ into **R** having the following properties:

   (i) $(x \mid y) = (y \mid x)$;

   (ii) $(x + x' \mid y) = (x \mid y) + (x' \mid y)$;

   (iii) $(\lambda x \mid y) = \lambda (x \mid y)$ for all $\lambda \in \mathbf{R}$;

   (iv) $(x \mid x) > 0$ save for $x = 0$.

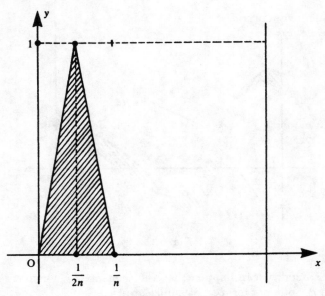

Fig. 44.

Such a map is again called a *scalar product* on E, and the structure that it defines on the vector space E is said to be the structure of a *prehilbert space*.

An example which immediately generalizes the ordinary space $\mathbf{R}^3$ is the vector space $\mathbf{R}^n$, where n is any natural number > 3, and where, for two vectors x and y, we set

$$(x \mid y) = \xi_1\eta_1 + \xi_2\eta_2 + \ldots + \xi_n\eta_n ,$$

where the ξ_j are the components of x and the η_j those of y (§1, B). The verification of Axioms (i) to (iv) is immediate. Another example is the vector space $C(0, 1)$ of real-valued, continuous functions on the interval $[0, 1]$ (Section A); a scalar product is here defined by

$$(6) \qquad\qquad (f \mid g) = \int_0^1 f(x)g(x)\,dx .$$

Axioms (i), (ii), and (iii) are verified in an obvious way. For (iv), it is necessary to prove that, if

$$(f \mid f) = \int_0^1 f^2(x)\,dx = 0 ,$$

then necessarily $f(x) = 0$ for all $x \in [0, 1]$. One argues as in Section A), considering the primitive

$$F(x) = \int_0^x f^2(t)\,dt .$$

The fundamental property of a scalar product is the *Cauchy-Buniakowski-Schwarz inequality*

(7) $$|(x\,|\,y)| \le \sqrt{(x\,|\,x)} \cdot \sqrt{(y\,|\,y)}\,,$$

in which equality only occurs if we have $\alpha x + \beta y = 0$ for two scalars α and β, not both zero.

Indeed, let us note that for *each* real number t, we have

$$(x + ty\,|\,x + ty) \ge 0\,,$$

which gives

(8) $$(x\,|\,x) + 2t\,(x\,|\,y) + t^2\,(y\,|\,y) \ge 0\,.$$

If $(y\,|\,y) = 0$ (that is to say, if $y = 0$), then we also have $(x\,|\,y) = 0$. Otherwise, let us give t the value

$$t_0 = -\frac{(x\,|\,y)}{(y\,|\,y)}$$

in (8). We obtain

(9) $$(x\,|\,x) - \frac{(x\,|\,y)^2}{(y\,|\,y)} \ge 0\,,$$

that is to say,

$$(x\,|\,y)^2 \le (x\,|\,x)(y\,|\,y)\,,$$

from which (7) follows on taking positive square roots. Further, if the left-hand side of (9) is zero, then this means that

$$(x + t_0 y\,|\,x + t_0 y) = 0\,,$$

that is to say, $x + t_0 y = 0$.

On setting $\|\,x\,\| = \sqrt{(x\,|\,x)}$, inequality (7) can be written

(10) $$|(x\,|\,y)| \le \|\,x\,\| \cdot \|\,y\,\|\,.$$

Minkowski's inequality follows:

(11) $$\|\,x + y\,\| \le \|\,x\,\| + \|\,y\,\|\,.$$

Indeed, we have

$$\begin{aligned}
\|\,x + y\,\|^2 &= (x + y\,|\,x + y) \\
&= (x\,|\,x) + 2\,(x\,|\,y) + (y\,|\,y) \\
&\le (x\,|\,x) + 2\,|(x\,|\,y)| + (y\,|\,y) \\
&\le \|\,x\,\|^2 + 2\,\|\,x\,\| \cdot \|\,y\,\| + \|\,y\,\|^2 \qquad \text{from (10)} \\
&= (\|\,x\,\| + \|\,y\,\|)^2\,,
\end{aligned}$$

from which (11) follows on taking square roots.

This inequality allows us to define a *distance* on the prehilbert space E by

(12) $$d(x, y) = \| x - y \|$$

because we can only have $d(x, y) = 0$ if $x = y$ by (iv). It is a consequence of (i) and (iii) that $(-z \mid -z) = (z \mid z)$, whence $d(y, x) = d(x, y)$. Finally we see from Minkowski's inequality that

$$\| x - z \| = \| (x - y) + (y - z) \| \leq \| x - y \| + \| y - z \| ,$$

that is to say, that $d(x, z) \leq d(x, y) + d(y, z)$. Of course, in this way we rediscover in $\mathbf{R}^3$ the inequality of a Euclidean triangle.

We see that we have also defined on $C(0, 1)$ a third distance

(13) $$d_2(f, g) = \sqrt{\int_0^1 (f(x) - g(x))^2 \, dx} .$$

It is easily shown that this distance is equivalent neither to the distance $d_\infty(f, g)$ nor to the distance $d_1(f, g)$.

C. Hilbert Spaces

We denote by ℓ^2 or $\ell^2_{\mathbf{R}}$ the set of *infinite sequences* of real numbers $x = (x_n)_{n=1,2,\ldots}$ such that the *series* with positive or zero terms

$$\sum_{n=1}^{\infty} x_n^2 = x_1^2 + x_2^2 + \ldots + x_n^2 + \ldots$$

is *convergent*. First a structure of a *vector space* over the field $\mathbf{R}$ (§3, F) is defined on this set. The *sum* of two sequences $x = (x_n)$ and $y = (y_n)$ of ℓ^2 is by definition the sequence

$$x + y = (x_n + y_n) ,$$

and it is necessary to show that the series

$$\sum_{n=1}^{\infty} (x_n + y_n)^2$$

is convergent. But this immediately follows from the inequality

$$(x_n + y_n)^2 \leq 2 \left(x_n^2 + y_n^2 \right) .$$

The *product* of $x = (x_n)$ by a *scalar* $\lambda \in \mathbf{R}$ is by definition

$$\lambda x = (\lambda x_n) ,$$

and, since $(\lambda x_n)^2 = \lambda^2 x_n^2$, it is clear that the series

$$\sum_{n=1}^{\infty} (\lambda x_n)^2$$

is convergent, and thus indeed $\lambda x \in \ell^2$. The verification of the axioms of vector spaces (§3, F) is immediate.

On this vector space, a *scalar product* is defined by

$$(x \mid y) = \sum_{n=1}^{\infty} x_n y_n \,.$$

It is clearly necessary to show that the series on the right-hand side is convergent. In fact, it is *absolutely convergent*, as a consequence of the inequality

$$|x_n y_n| \leq \frac{1}{2} \left(x_n^2 + y_n^2 \right) \,.$$

Axioms (i), (ii), and (iii) of Section B) are verified in an obvious way. As to (iv), to show that, if $(x \mid x) = 0$, that is to say, if

$$x_1^2 + x_2^2 + \ldots + x_n^2 + \ldots = 0 \,,$$

then each of the x_j is zero, it is sufficient to note that

$$0 \leq x_j^2 \leq x_1^2 + x_2^2 + \ldots + x_n^2 + \ldots$$

for each j.

The space $\ell_{\mathbf{R}}^2$, furnished with the scalar product (14), is said to be a *Hilbert space* (or a *real* Hilbert space).

D. p-adic Distances

Let p be a prime number.

Each non-zero integer m can be written

$$m = p^{\alpha} m' \,,$$

where the integer m' is not divisible by p and α is a non-negative integer. The number α is said to be the *p-adic valuation* of m, and it is denoted by $v_p(m)$. We thus have:

(15) $v_p(m) = 0$ if p does not divide m;

(16) $v_p(-m) = v_p(m)$;

(17) $v_p(mn) = v_p(m) + v_p(n)$.

Finally, we have the inequality

(18) $v_p(m + n) \geq \min\{v_p(m), v_p(n)\}$ if $m + n \neq 0$,

the right-hand side being the smaller of the two numbers $v_p(m)$ and $v_p(n)$. To prove this, let us write $m = p^\alpha m'$ and $n = p^\beta n'$, where p divides neither m' nor n'. For example, let us suppose that $\alpha \geq \beta$. Then

$$m + n = p^\beta (p^{\alpha-\beta} m' + n').$$

Since it may happen that p divides $p^{\alpha-\beta} m' + n'$ in the case where $\alpha = \beta$, we can only say that $v_p(m+n) \geq \beta$. However, if $\alpha > \beta$, then we have the equality

$$v_p(m + n) = \min\{v_p(m), v_p(n)\}.$$

Let us now define for *each* integer m, the *absolute p-adic value* $|m|_p$. If $m = 0$, then we set $|0|_p = 0$. Otherwise, we take

(19) $$|m|_p = p^{-v_p(m)}.$$

From the relations (15), (16), and (17), it then follows that:

(20) $\quad |m|_p = 1$ if $m \neq 0$ and p does not divide m;

(21) $\quad |-m|_p = |m|_p$;

(22) $\quad |mn|_p = |m|_p \cdot |n|_p$.

But the relations (21) and (22) are clearly also true if $m = 0$ or if $n = 0$. Finally, for *all* integers m and n, we have

(23) $$|m + n|_p \leq \max(|m|_p, |n|_p) \leq |m|_p + |n|_p.$$

This is in fact evident by definition if $m = 0$ or if $n = 0$ or if $m + n = 0$. Otherwise, it is a consequence of (18).

Finally, a distance can be defined on the set $\mathbf{Z}$ of all integers by the formula

(24) $$d_p(m, n) = |m - n|_p.$$

The verification of the axioms for distances in fact follows immediately from the definitions and the relations (21) and (23). We say that $d_p(m, n)$ is the *p-adic distance* from m to n.

This distance has properties which are very different from those of the usual distance $|m - n|$, and which can seem very strange. If a non-zero integer a is not divisible by p, then we have $|a^n|_p = 1$ for *all* $n \geq 0$. On the other hand, $|p^n|_p = p^{-n}$ *converges to* 0 when n tends to $+\infty$. This also shows that for two different prime numbers p and q, the distances d_p and d_q are not equivalent.

5. Fourier Series

A. Trigonometric Series and Fourier Coefficients

We shall suppose known two fundamental properties of series of continuous functions on an interval $[a, b]$:

(1) $$f_0(x) + f_1(x) + \ldots + f_n(x) + \ldots.$$

(i) If $A_0 + A_1 + \ldots + A_n + \ldots$ is a convergent series of numbers $A_n \geq 0$, and if $|f_n(x)| \leq A_n$ for all $x \in [a, b]$, then the series (1) is uniformly convergent, and the sum $f(x)$ is *continuous* on $[a, b]$.

(ii) Under the same hypotheses, we have

$$(2) \quad \int_a^b f(x)\,dx = \int_a^b f_0(x)\,dx + \int_a^b f_1(x)\,dx + \ldots + \int_a^b f_n(x)\,dx + \ldots,$$

where the series on the right-hand side is absolutely convergent, and

$$\int_a^b |f(x)|\,dx \leq (b-a)(A_0 + A_1 + \ldots + A_n + \ldots).$$

A series of the form

$$(3) \quad a_0 + (a_1 \cos 2\pi x + b_1 \sin 2\pi x) + \ldots$$
$$+ (a_n \cos 2\pi n x + b_n \sin 2\pi n x) + \ldots$$

is called a *trigonometric series* of period 1. If this series converges at a point $x \in \mathbf{R}$, then it converges at all the points $x + m$, where m is an integer, with the same sum. If it converges at all of the points $x \in [0, 1]$, it thus converges on $\mathbf{R}$, and its sum $f(x)$ is *periodic* of period 1, in other words $f(x + 1) = f(x)$.

We are going to consider in particular the case where the series

$$(4) \quad |a_0| + (|a_1| + |b_1|) + (|a_2| + |b_2|) + \ldots + (|a_n| + |b_n|) + \ldots$$

is convergent. Since $|\cos 2\pi n x| \leq 1$ and $|\sin 2\pi n x| \leq 1$, the series (3) is thus absolutely and uniformly convergent, and its sum $f(x)$ is continuous on $\mathbf{R}$ and periodic of period 1. Further, it is noteworthy that one can *express the coefficients a_k and b_k with the aid of $f(x)$*. This is a consequence of the following so-called *orthogonality* formulae:

for all integers m and n,

$$(5) \quad \int_0^1 \cos 2\pi m x \cdot \sin 2\pi n x\,dx = 0;$$

for integers m and n with $m \neq n$,

$$(6) \quad \int_0^1 \cos 2\pi m x \cdot \cos 2\pi n x\,dx = 0, \quad \int_0^1 \sin 2\pi m x \cdot \sin 2\pi n x\,dx = 0;$$

and for $m \neq 0$,

$$(7) \quad \int_0^1 \cos^2 2\pi m x\,dx = \frac{1}{2}, \quad \int_0^1 \sin^2 2\pi m x\,dx = \frac{1}{2}.$$

These formulae are deduced immediately from the classic trigonometric formulae

$$\cos a \cdot \sin b = \frac{1}{2}\left(\sin(a+b) + \sin(b-a)\right)$$

$$\cos a \cdot \cos b = \frac{1}{2}\left(\cos(a+b) + \cos(a-b)\right)$$

$$\sin a \cdot \sin b = \frac{1}{2}\left(\cos(a+b) - \cos(a-b)\right)$$

and from the fact that, for each non-zero integer p,

$$\int_0^1 \cos 2\pi px\, dx = 0\,,$$

$$\int_0^1 \sin 2\pi px\, dx = 0\,.$$

Under the hypothesis of the convergence of the series (4), we thus have *Fourier's formulae* giving the coefficients of (3) with the aid of $f(x)$:

$$a_0 = \int_0^1 f(x)\, dx\,,$$

(8) $\qquad a_n = 2\int_0^1 f(x)\cos 2\pi nx\, dx\,,$

$$b_n = 2\int_0^1 f(x)\sin 2\pi nx\, dx \quad \text{for} \quad n \geq 1\,.$$

In fact, as a consequence of property (ii), to calculate the integral

$$\int_0^1 f(x)\cos 2\pi nx\, dx\,,$$

the series (3) can be multiplied by $\cos 2\pi nx$ and integrated term-by-term; the result follows from formulae (5), (6), and (7). The integral

$$\int_0^1 f(x)\sin 2\pi nx\, dx$$

is calculated in the same way.

For *every* continuous function $f(x)$ on $[0,1]$ (but which is not supposed to be the sum of a series (3)), the coefficients a_0, a_n, and b_n for $n \geq 1$ defined by the formulae (8) are called the *Fourier coefficients* of f.

When f has a *continuous derivative* f', the Fourier coefficients of f',

$$a_0' = \int_0^1 f'(x)\, dx\,,$$

(9) $\qquad a_n' = 2\int_0^1 f'(x)\cos 2\pi nx\, dx\,,$

$$b_n' = 2\int_0^1 f'(x)\sin 2\pi nx\, dx \quad \text{for} \quad n \geq 1\,,$$

are also defined, and we have the following relations between these coefficients and those of f:

$$
\begin{aligned}
a_0' &= f(1) - f(0)\,,\\
a_n &= -\frac{1}{2\pi n} b_n'\,,\\
b_n &= \frac{1}{2\pi n}(a_n' - a_0') \quad \text{for} \quad n \geq 1\,.
\end{aligned}
$$

(10)

These are proved simply by integration by parts, for example

$$
\begin{aligned}
a_n &= 2 \int_0^1 f'(x) \cos 2\pi nx\, dx\\
&= \left[\frac{1}{\pi n} f(x) \sin 2\pi nx\right]_0^1 - \frac{1}{\pi n} \int_0^1 f'(x) \sin 2\pi nx\, dx\,.
\end{aligned}
$$

B. The Convergence of Fourier Series

For *every* continuous function f on $[0,1]$, once we have formed the Fourier coefficients (8) of f, we can consider the corresponding trigonometric series (3). We say that this is the *Fourier series* of f. Two problems are immediately posed : does this series converge, and, when it does converge at a point x, is its sum $f(x)$?

When it is *only* assumed that f is continuous, these are very difficult problems. One had to wait until 1873 before an example of a continuous function whose Fourier series *does not converge* at certain points was found, and it was not until 1966 that it became possible to show that these points form only a set which is in a certain sense "negligible", the series being convergent and having sum $f(x)$ at the other points.

We shall limit ourselves to the case where the series (4), whose terms are the absolute values of the Fourier coefficients of f, is convergent. In this case, the Fourier series (3) converges for each $x \in \mathbb{R}$, as we saw in A), and has for its sum a continuous, periodic function $g(x)$. It *remains to show that* $g = f$.

We use an auxiliary series which arises directly from the geometric series

(11)
$$
\frac{1}{1-z} = 1 + z + z^2 + \ldots + z^n + \ldots\,,
$$

valid for all complex numbers z such that $r = |z| < 1$. If we set

$$
z = r\,(\cos t + i \sin t)\,,
$$

then we have

$$
z^n = r^n\,(\cos nt + i \sin nt)
$$

(a formula proved by Cotes and de Moivre at the beginning of the eighteenth century). On taking the real parts of the two sides of (11), we obtain the relation

(12) $\dfrac{1-r^2}{1+r^2-2r\cos t} = 1 + 2\,(r\cos t + r^2\cos 2t + \ldots + r^n\cos nt + \ldots)$,

the series on the right-hand side being (for fixed $r < 1$) uniformly convergent on **R**. We set, for $0 < r < 1$,

(13) $\Phi_r(x) = \dfrac{1-r^2}{1+r^2-2r\cos 2\pi x} = 1 + 2\displaystyle\sum_{n=1}^{\infty} r^n\cos 2\pi nx$.

This is a periodic function of period 1, continuous on **R**, which has the following three properties:

(i) $\Phi_r(x) > 0$ for all $x \in$ **R**. The denominator in (13) is in fact none other than

$$|1 - r\,(\cos 2\pi x + i\sin 2\pi x)|^2$$

as a consequence of the calculation done above starting from (11).

(ii) For each number δ such that $0 < \delta < 1/2$, and for all x such that $\delta \le x \le 1 - \delta$, we have $\cos 2\pi x \le \cos 2\pi\delta$, and hence

$$\begin{aligned}
1 + r^2 - 2r\cos 2\pi x &\ge 1 + r^2 - 2r\cos 2\pi\delta \\
&= (1 - r^2) + 4r\sin^2\pi\delta \\
&\ge 4r\sin^2\pi\delta.
\end{aligned}$$

Thus, if $\delta \le x \le 1 - \delta$, then we have

(14) $\Phi_r(x) \le \dfrac{1-r^2}{4r\sin^2\pi\delta} \le \dfrac{1-r}{2r\sin^2\pi\delta}$.

(iii)
$$\int_0^1 \Phi_r(x)\,dx = 1.$$

In fact, the series of the right-hand side of (13) can be integrated term-by-term, and all the integrals are zero except the first.

Having established these results, let us form the following function:

(15) $F_r(x) = \displaystyle\int_0^1 \Phi_r(x-t)f(t)\,dt$.

We have, for all $x \in$ **R** and all $t \in [0, 1]$,

(16) $\Phi_r(x-t)f(t) = f(t) + 2\displaystyle\sum_{n=1}^{\infty} r^n f(t)\cos 2\pi n(x-t)$.

Since $|f(t)| \le M^2$, we have

$$|r^n f(t)\cos 2\pi n(x-t)| \le Mr^n,$$

---

[2] All continous functions on $[0, 1]$ are bounded.

so that the series on the right-hand side of (16) is uniformly convergent and can be integrated term-by-term. We obtain

$$(17) \qquad F_r(x) = \int_0^1 f(t)\,dt + 2\sum_{n=1}^{\infty} r^n \int_0^1 f(t)\,\cos 2\pi n(x-t)\,dt.$$

Since

$$\cos 2\pi n(x-t) = \cos 2\pi nx \cdot \cos 2\pi nt + \sin 2\pi nx \cdot \sin 2\pi nt,$$

we have

$$2\int_0^1 f(t)\,\cos 2\pi n(x-t)\,dt = a_n \cos 2\pi nx + b_n \sin 2\pi nx$$

from (8), whence, finally,

$$(18) \qquad F_r(x) = a_0 + \sum_{n=1}^{\infty} r^n (a_n \cos 2\pi nx + b_n \sin 2\pi nx).$$

The relation

$$(19) \qquad f(x) = a_0 + \sum_{n=1}^{\infty} (a_n \cos 2\pi nx + b_n \sin 2\pi nx)$$

will thus be established if it is proved that:

1) When r tends to 1, the series on the right-hand side of (18) tends to the Fourier series of f.

2) When r tends to 1, $F_r(x)$ tends to $f(x)$ for all $x \in [0,1]$.

Proof of 1). The difference between the series on the right-hand sides of (18) and (19) is

$$(20) \qquad \sum_{n=1}^{\infty} (1 - r^n)(a_n \cos 2\pi nx + b_n \sin 2\pi nx).$$

Given a number $\varepsilon > 0$, there is a natural number N such that

$$\sum_{n=N+1}^{\infty} (|a_n| + |b_n|) \le \varepsilon,$$

and consequently

$$(21) \qquad \left| \sum_{n=N+1}^{\infty} (1 - r^n)(a_n \cos 2\pi nx + b_n \sin 2\pi nx) \right| \le 2\varepsilon.$$

On the other hand, for $1 \le k \le N$, we have

$$1 - r^k = (1 - r)(1 + r + r^2 + \ldots + r^{k-1}) \le N(1 - r).$$

If we set

$$A = \sum_{n=1}^{\infty} (|a_n| + |b_n|) \, ,$$

we thus have

$$(22) \qquad \left| \sum_{n=1}^{N} (1 - r^n)(a_n \cos 2\pi nx + b_n \sin 2\pi nx) \right| \leq AN(1 - r),$$

and so it follows from (21) and (22) that we have, for the series (20),

$$(23) \qquad \left| \sum_{n=1}^{\infty} (1 - r^n)(a_n \cos 2\pi nx + b_n \sin 2\pi nx) \right| \leq AN(1 - r) + 2\varepsilon \, .$$

Thus, if r is sufficiently close to 1 that $1 - r \leq \varepsilon/AN$, then the right-hand side of (23) is bounded by 3ε, which concludes the proof.

Proof of 2). Let us fix a number $x \in [0, 1]$, and let us suppose given an $\varepsilon > 0$. Since f is continuous, there is a number δ such that $0 < \delta < 1/2$ and such that we have

$$(24) \qquad |f(x) - f(t)| \leq \varepsilon \quad \text{for} \quad x - \delta \leq t \leq x + \delta \, .$$

(If $x - \delta < 0$ or $x + \delta > 1$, it is necessary to replace $x - \delta$ by 0 or $x + \delta$ by 1, respectively, in this condition.) Notice now that, as a consequence of property (iii) of $\Phi_r(x)$, we can write

$$(25) \qquad f(x) = \int_0^1 \Phi_r(x - t) f(x) \, dt \, .$$

To see this, we use the fact that for each continuous function h on **R** which is periodic of period 1, we have

$$\int_0^1 h(t) \, dt = \int_a^{a+1} h(t) \, dt$$

for each $a \in \mathbf{R}$. If $a = n$ is an integer, it suffices to note that the change of variable $t = n + u$ gives

$$\int_n^{n+1} h(t) \, dt = \int_0^1 h(n + u) \, du = \int_0^1 h(u) \, du$$

by periodicity. If $n < a < n + 1$ for some integer n, then we have

$$\int_a^{a+1} h(t) \, dt = \int_a^{n+1} h(t) \, dt + \int_{n+1}^{a+1} h(t) \, dt \, ,$$

and, by the change of variable $t = y + 1$, we have

$$\int_{n+1}^{a+1} h(t) \, dt = \int_n^a h(y + 1) \, dy = \int_n^a h(t) \, dt \, ,$$

giving (25).

From (25), we obtain

$$f(x) - F_r(x) = \int_0^1 \Phi_r(x - t)(f(x) - f(t))\, dt\,.$$

To simplify, we write $g(t, x) = \Phi_r(x - t)(f(x) - f(t))$. Then we have

$$(26) \qquad \int_0^1 g(t, x)\, dt = \int_0^{x-\delta} g(t, x)\, dt + \int_{x-\delta}^{x+\delta} g(t, x)\, dt + \int_{x+\delta}^1 g(t, x)\, dt\,.$$

(If $x - \delta < 0$ or $x + \delta > 1$, then we suppress the first and third integrals, respectively, and we replace $x - \delta$ by 0 and $x + \delta$ by 1, respectively, in the second one.)

We can separately majorize each of the three integrals in (26). We know that $\Phi_r(x - t) > 0$, and so it follows from (24) that

$$\left| \int_{x-\delta}^{x+\delta} \Phi_r(x - t)(f(x) - f(t))\, dt \right| \le \varepsilon \int_{x-\delta}^{x+\delta} \Phi_r(x - t)\, dt \le \varepsilon$$

because

$$\int_{x-\delta}^{x+\delta} \Phi_r(x - t)\, dt \le \int_0^1 \Phi_r(x - t)\, dt = 1\,.$$

On the other hand, as a consequence of (14),

$$\left| \int_0^{x-\delta} \Phi_r(x - t)(f(x) - f(t))\, dt \right| \le 2M \frac{1 - r}{2r \sin^2 \pi\delta}\,,$$

and we have the same majorization for the integral extending from $x + \delta$ to 1.

Finally, we have the inequality

$$(27) \qquad |f(x) - F_r(x)| \le \varepsilon + 2M \frac{1 - r}{r \sin^2 \pi\delta} \le \varepsilon + 4M \frac{1 - r}{\sin^2 \pi\delta}$$

if $r > 1/2$. If then r is so close to 1 that $r > 1/2$ and

$$1 - r \le \frac{\varepsilon \sin^2 \pi\delta}{4M}\,,$$

then $|f(x) - F_r(x)| \le 2\varepsilon$, and (19) is proved.

C. Fourier Series of Bernoulli Polynomials

Recall (Chapter IV, Appendix 4) that for each natural number n, we have defined the *Bernoulli polynomial* $\varphi_n(x)$ of degree n. We recall the principal properties:

$$(28) \qquad \varphi_{2k+1}(0) = \varphi_{2k+1}(1) = 0, \quad \varphi_{2k}(0) = \varphi_{2k}(1) = (-1)^{k+1} B_k$$

for $k \geq 1$;

(29)
$$\varphi_n'(x) = n\varphi_{n-1}(x)$$

for $n \geq 2$. Finally, we have

$$\varphi_1(x) = x - \frac{1}{2},$$

so that $\varphi_1'(x) = 1$.

These formulae and the formulae (10) immediately determine the *Fourier coefficients* of the φ_n. For φ_1, we have

(30)
$$a_n = 0 \quad \text{for} \quad n \geq 0, \qquad b_n = -\frac{1}{\pi n}.$$

For all the φ_n, we have

$$a_0 = \int_0^1 \varphi_n(x)\,dx = 0,$$

because this can be written

$$\frac{1}{n+1} \int_0^1 \varphi_{n+1}'(x)\,dx = \frac{1}{n+1}(\varphi_{n+1}(1) - \varphi_{n+1}(0)) = 0$$

by (28).

The other Fourier coefficients are deduced from (30) and the formulae (10) by recurrence on n. For $\varphi_{2k}(x)$, we have

$$a_n = (-1)^{k+1} \frac{2\,(2k)!}{(2\pi n)^{2k}}, \qquad b_n = 0,$$

and for $\varphi_{2k+1}(x)$, we have

$$a_n = 0, \qquad b_n = (-1)^{k+1} \frac{2\,(2k+1)!}{(2\pi n)^{2k+1}}.$$

Since, for all the polynomials $\varphi_m(x)$ starting from $m \geq 2$, the series

$$\sum_{n=1}^{\infty} (|a_n| + |b_n|)$$

converges, we easily obtain equations (14) and (15) of Chapter IV, Appendix 4.

D. Cantor's Problems

It can happen that a trigonometric series (3) is *convergent* (but not absolutely convergent) for all $x \in \mathbf{R}$. But its sum is not necessarily continuous, and the integrals (8) may not have any meaning. This is, for example, the case with the series

$$\sum_{n=1}^{\infty} \frac{1}{\sqrt{\log n}} \sin 2\pi n x \,.$$

The concept of a trigonometric series is thus more general than that of a Fourier series. It was first considered by Riemann, who could prove the following theorem: if the series (3) converges for all x and has 0 for its sum, then all the coefficients a_n and b_n are zero. Cantor wondered if this conclusion were still true when it was supposed that the series (3) converges and has 0 for its sum *save at the points of a set E*. It was this that brought him to the study of arbitrary subsets of **R**, in particular from the point of view of their order structure or their topology. But he quickly abandoned his initial problem, which is still not completely resolved.

VI Problems and Pseudo-Problems about "Foundations"

I think I can say that what I have expounded in the last three chapters carries out the programme which I outlined in the introduction: to show how the "abstract" nature of mathematical objects of our day flows from methods invented between 1800 and 1930 with a view to the solving of *classical* problems; and to induce an awareness — restricted by my not being able to go into too much technical detail — that the usefulness of these methods and of the objects to which they are applied is greater today than it ever was.

We began with the "hypothetico-deductive" conception of mathematics already described by Plato and put into practice by Euclid in his *Elements*. Then, leaping, so to say, over the centuries, we saw how Pasch and Hilbert gave this conception a totally precise form for geometry, a form which was almost at once extended to all branches of mathematics by their contemporaries and their immediate successors, to become what can be called the *routine* of today's mathematicians.

But in doing this we passed in silence over a whole aspect of this same period 1800 — 1930 : an epoch which, as we have emphasized, was one of extraordinary fruitfulness in the creation of new theories, but which was also one in which doubts and controversies about the nature of mathematics were at their height. The debates which arose from these are scarcely of interest now except to historians and philosophers. Nevertheless, in the same way as the controversies of the seventeenth and eighteenth centuries about the "infinitely small", they left a positive gain in the increased precision of mathematical language, and in particular they made way for enormous advances in mathematical logic in the twentieth century.

As we said at the beginning of Chapter V, it was mathematicians' state of mind as regards the nature of mathematical objects which changed after 1800. Instead of resting content, as their predecessors had done, with the giant strides which could be made using the tools forged in the seventeenth century, and thinking that this was enough to confirm the truth of the principles on which these methods were based, many began to ask questions about the very meaning of these principles whose "truth" was being proclaimed. If I wished to sum up in one sentence the way in which ideas unfolded during this period, I would say that its essence was a progressive abandonment of the concept of "evident truths", first in geometry, and then in the rest of mathematics.

1. Non-Euclidean Geometries

A. The Parallel Postulate

We have said (Chapter III, 4) that the attitude of mathematicians of the seventeenth and eighteenth centuries regarding geometrical postulates was very different from that of Plato and Euclid. While the latter spoke only of "hypotheses" or "requirements", the former wanted to see in them "truths", without explaining how this characteristic is to be seen in concepts "not existing in nature", as Lobachevski said; rightly complaining of the obscurity of the "words with which we begin to do geometry". To say, with Kant, that these postulates are "synthetic *a priori* judgements" may satisfy philosophers, but is of no help to a mathematician in giving precision to his ideas.

There is thus a general consensus in accepting the perfect harmony between the properties of the geometric objects described by the postulates and the behaviour of their "images" in nature. Nobody, for example, seems to find the assertion that a straight line can be indefinitely extended anything but obvious: perhaps because this corresponds to what one can do by way of experiment, "approximately" and within very restricted limits[1].

There was nevertheless one "requirement" of Euclid which even in classical times was the subject of discussion: this is the famous "parallel postulate", which eventually was to be called *the* "Euclidean postulate". Given a point A in a plane, not on a line Δ, it is easy to draw *one* line Δ' passing through A parallel to Δ: it is only necessary (Figure 45) to draw the line AB perpendicular to Δ, then the line Δ' perpendicular to AB at the point A. The line Δ' cannot meet Δ, for, if it did, two distinct lines both perpendicular to AB would pass through the common point of Δ and Δ', contradicting a previous theorem (Euclid, *Elements*, Book I, 27).

What the postulate affirms is that Δ' is the *only* line parallel to Δ which passes through A. At least, that is how it is stated today. Euclid presented this property in a more positive form: if, in a half-plane formed from points on one side of a line D, two half-lines AX_1 and BX_2 form angles α and β, respectively, with the segment AB of D, whose sum is less than two right angles (Figure 46), then these two half-lines have a common point (*Elements*, Book I, Postulate 5).

The uniqueness of the line parallel to Δ passing through A in Figure 45 is deduced from the above postulate: a line $Y''X''$ passing through A and such that the angle $\widehat{BAX''}$ is less than a right angle meets BX, because the sum of the angles $\widehat{BAX''}$ and $\widehat{ABX}$ is less than two right angles. The argument is

---

[1] Perhaps also, although the postulates themselves have no convincing "crude" images, their "truth" was accepted without difficulty because their logical consequences provided models in natural science, according well with experience. The same situation occurred with respect to Newton's laws for the dynamics of "material points": although their existence in themselves was more or less impossible to prove, it was possible to deduce from them a host of consequences giving altogether satisfying "explanations" of phenomena.

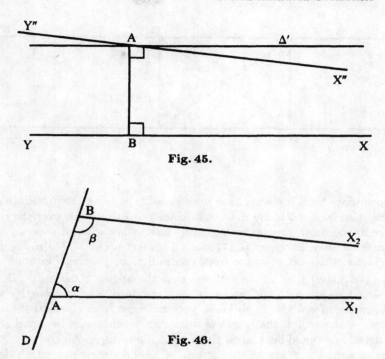

Fig. 45.

Fig. 46.

the same if the angle $\widehat{BAX''}$ is greater than a right angle, with respect to the half-lines AY'' and BY.

Although he is still dealing with an affirmation which is on a par with the indefinite extension of straight lines, since it is a matter of accepting the existence of a point at an arbitrarily large distance, Euclid could not do without it to establish his principal theorems, for example the existence of *rectangles*, which everybody is inclined to take as a "truth". In fact (Figure 47), in order to complete a rectangle with three corners A, B, C already fixed, with $\widehat{ABC}$ equal to a right angle, a segment CD equal in length to AB is taken from the line perpendicular to BC at the point C, and then it has to be proved that the angles $\widehat{BAD}$ and $\widehat{CDA}$ are right angles. Since the angles $\widehat{ABC}$ and $\widehat{BCD}$ are equal, and since $AB = CD$, the triangles ABC and BCD are congruent, and therefore $AC = BD$. The corresponding sides of the triangles BAD and CDA being equal, the angles $\widehat{BAD}$ and $\widehat{CDA}$ are equal; but *why should these angles be right angles*? If the postulate be accepted, the argument is as follows: if the angle $\widehat{BAD}$ is acute, the half-lines AX' and BX meet, and since the angle $\widehat{CDA}$ is also acute, the half-lines DY' and CY meet as well; the two lines XY and $X'Y'$ would then have two common points, which is absurd. If the angle $\widehat{BAD}$ is obtuse, the angles $\widehat{BAY'}$ and $\widehat{CDX'}$ are acute, and the same argument again leads to a contradiction.

The objections to this proof, raised since classical times, are not concerned with its validity, but with a failure to understand why Euclid had to introduce

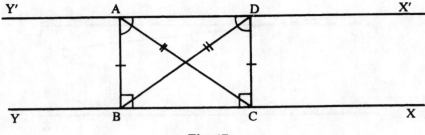

Fig. 47.

a new postulate here. It should have been possible to prove the postulate using only the other axioms. Many mathematicians tried to do this over the course of the centuries, from Proclus to Legendre, but always without success. In the eighteenth century, Saccheri in 1733 and Lambert around 1770 inaugurated a new tactic: this was to argue from contradiction, proving (without using Euclid's postulate) that the hypothesis that the angles $\widehat{BAD}$ and $\widehat{CDA}$ are acute[2] leads to absurd conclusions. They developed a whole lengthy series of consequences from this hypothesis, very different from classical theorems: for example, the sum of the angles of a triangle would never be equal to two right angles, but would depend on the *area* of the triangle. All the same, none of these strange theorems contradict any of Euclid's theorems which do not depend on the postulate.

During the years 1795 — 1830, these repeated failures had the eventual result of persuading several mathematicians that the contradiction which they sought would *never* be reached, and that therefore there was room in mathematics for *several* different geometries. But this belief was in complete opposition to received opinion in the intellectual circles of the time, which invested Euclidean geometry with a quality of necessity which was inseparable from our concept of space. Gauss, who from his youth took an interest in the problem of parallels, had, by about 1816, hit upon the idea that it is possible, if we accept as a "hypothesis" (in Plato's sense) the *negation* of Euclid's postulate, to found a geometry which he called "non-Euclidean", and which he judged to be, despite its strangeness, "entirely consistent with itself". He shared his reflections with a few friends, but refused to publish them, replying to those who urged him to do so: "I am too much afraid of the outcries of the Boeotians if I declared my views in full".

A little later, two other mathematicians, the Hungarian J. Bolyai and the Russian N. Lobachevski, arrived independently at Gauss's conclusions. Each published, in about 1830, a treatise on "non-Euclidean" geometry, developed according to the Euclidean model. Until about 1860, however, these works remained unknown to most mathematicians, and those who knew of them

---

[2] It can be shown that the hypothesis that these angles are obtuse contradicts Archimedes' axiom (see Chapter III, Appendix 2).

judged them only as fantasies without any serious implications, no true part of mathematics.

B. Geometry on a Surface

The situation changed after 1860, under the influence of differential geometry. Inspired by his many works on geodesics and astronomy, Gauss was no doubt the first to conceive of the idea that it is possible to construct a "geometry" on *any surface* Σ that is sufficiently regular[3], once it is known how to define the *length* of an arc of a sufficiently regular curve drawn on Σ. One can in fact define here the objects which correspond to the straight lines of the Euclidean plane, and which are called the *geodesics* of Σ: such a curve Γ is characterized by the property that for two points p and q of Γ, *sufficiently close together*, the arc of Γ between p and q is the *shortest* of any arcs of a curve *on Σ* with endpoints p and q. For example, on a sphere the geodesics are the *great circles*, intersections of the sphere with planes passing through the centre. On a right cylinder with a circular base, the geodesics are :

1) the (vertical) generators;

2) circles cutting across the cylinder on planes perpendicular to the axis;

3) all the *helixes* which can be drawn on the cylinder.

Knowing the lengths of the curves drawn on Σ enables us firstly to define the *angle* at which two geodesics intersect at a point, and in particular to define the *right* angle (Appendix 1). But, in trying to develop a "geometry" on Σ following Euclid's model, we nearly always have to limit ourselves to a *small* domain, for it is only in a sufficiently small set Δ that there exists *one and only one* geodesic arc joining two arbitrary points p and q of Δ and *completely contained in Δ*. Effectively, then, the geodesic specifies the smallest value of the length of an arc of a curve on Σ joining p and q[4]. Thus, in Δ, the first of Euclid's "requirements" (see Chapter III, 4) is satisfied; and it is the same with the second, the "extension" of a segment of geodesic, so long as we do not set out to make this extension "indefinite"[5]. The *"areas"* of sufficiently regular figures contained in Δ can also be defined.

In general, however, the "geometry" on Δ must stop at this point. There are no longer "cases of congruence of triangles"[6]: for example, two triangles in Δ can have corresponding sides of the same length, without the corresponding angles being equal. This is because in general we cannot have *displacements* in Δ, that is to say, transformations which keep geodesics, lengths, and angles the same.

---

[3] For a precise definition, see Appendix 1.

[4] On a circular cylinder, there is always an *infinite number* of geodesics passing through two distinct points p and q, unless these points are on a circle which is a section of the cylinder by a plane perpendicular to the axis.

[5] In most of Euclid's propositions, there is no need to extend a line for more than a length which scarcely exceeds those of the segments in the "figure" being studied.

[6] The "triangles" concerned always have sides which are arcs of geodesics.

Nevertheless, on *certain* surfaces (other than the plane), such transformations do indeed exist. They have the property of *transitivity* , which means that an arc pq of a geodesic in Δ can be transformed into *any* other arc of a geodesic $p'q'$ of the *same length* by means of one of these transformations. This is what happens for example in the case of a sphere, where the "displacements" are rotations around an axis passing through the centre of the sphere; and there are still valid "cases of congruence" of triangles with sides of length less than half a great circle. This is also true for a circular cylinder, so long as Δ is taken to be, for example, the part projected on an arc of the base circle which is not the entire circle.

Here, however, we note a fundamental difference between the sphere and the cylinder: for the latter, the sum $A + B + C$ of the angles of a triangle ABC is still equal to two right angles; but on the sphere this sum is always *greater* than two right angles, no matter how small the triangle may be. To be precise, we have (for angles expressed in radians)

$$A + B + C - \pi = \frac{\text{area}(ABC)}{R^2} ,$$

where R is the radius of the sphere.

All this has been known in substance since classical times[7], but it does not seem to have had any influence on the problem of parallels, doubtless because neither the sphere nor the cylinder have the property that the arcs of geodesics joining two *arbitrary* points are *unique*, and because there are no *groups* of "displacements" of a small domain Δ[8].

C. Models of Non-Euclidean Geometry

The Greeks do not seem to have tried to calculate the length of any plane curve other than the circle. It was not until the seventeenth century that it became possible, by means of the infinitesimal calculus, to provide a formula enabling one to compute the length of a sufficiently regular plane curve: it is defined, as is the circle, by the limit of the length of a polygon "inscribed" in the circle, when the sides have a length tending to 0 (Figure 48).

Euler extended this formula to "skew " curves embedded in ordinary space $\mathbf{R}^3$ (Appendix 1): when he and his successors before Riemann spoke of the length of a curve drawn on a surface Σ, they disregarded the surface and considered the curve directly as embedded in $\mathbf{R}^3$ in order to calculate its length.

Gauss's researches, however, showed that, if the surface Σ can be "deformed" while keeping the curves drawn on Σ the same length (for example, "deforming" the base curve of a right cylinder without changing its length), then the "geometry" of the "deformed" surface must be considered as the same

---

[7] Geometry and trigonometry on a sphere were indispensable for calculating the positions of heavenly bodies

[8] When "displacements" are "composed", it is usual to move out of the domain Δ.

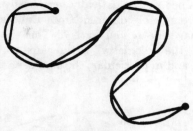

Fig. 48.

as on the original surface; or in other words, the embedding of the surface in $\mathbf{R}^3$ no longer comes into the calculations.

It was only after 1860 that anyone thought of the possibility of breaking free from this uniform procedure for defining lengths, and inventing many others. Let us consider for example a hemisphere Σ bounded by a plane disc Δ which contains its centre. Each point of Δ is the orthogonal projection of a unique point of the hemisphere, its "preimage" (Figure 49).

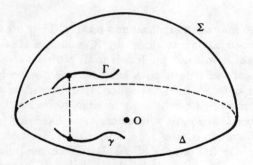

Fig. 49.

It can then be decreed that the "length" of a plane curve γ contained in Δ should be the length (in Euler's sense) of its preimage Γ, and all the arguments of Gauss are valid for this new definition of "length". But what is obtained in Δ is quite a different geometry from Euclidean geometry: the geodesics are the orthogonal projections of the half great circles of Σ, that is to say, the semi-ellipses whose major axis is a diameter of Δ, and the sum of the "angles" of a "triangle" is always greater than two right angles.

Another example is the definition of lengths on what is known as the "flat torus". A "crude" image of a torus is made by taking a circular cylinder of height h, and "bending" it so that the two circles at its ends are joined together (Figure 50). This surface can be mathematically defined without difficulty (Appendix 1), but here the "length" of a curve γ is taken to be the length (in Euler's sense) of the corresponding curve on the cylinder *before* it was "bent". This is defined mathematically very easily (Appendix 1), but it results in a definition completely different from Euler's; for example, all the parallels of the torus have the *same* length h! This time, on a small domain contained on the torus, the "geometry" is the same as for the corresponding domain of the cylinder, and in particular, the sum of the "angles" of a triangle is always equal to two right angles.

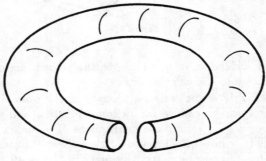

Fig. 50.

These examples may seem artificial and bizarre. They are only very special cases of general conceptions introduced by Riemann in his famous *Inaugural Lecture* of 1854, which was not published until after his death, in 1868. For Riemann, the concept of length on a surface is subject only to a very few conditions of an analytic nature (Appendix 1), and his ideas go much further, since he defines "lengths" straight away in what he calls a "multiplicity in n dimensions", where n is any natural number. As we have said (see Chapter V, 5 A, XVII), these "Riemann spaces", so foreign to our intuition, were eventually to play a major role in the physics of our day. From the point of view of our present concerns, it must obviously be stressed that Riemann could not have been interested in claiming "evident truth" for his axioms; and the title of his inaugural lecture mentions only "hypotheses set at the foundation of the geometry", as in Plato.

It was this way of thinking which enabled Beltrami [9] and Klein, in about 1870, finally to dissipate any doubts which might yet remain about the validity of "non-Euclidean" geometries. It could always have been objected against the "convictions" of the originators of these geometries that if they had pushed their arguments further, they could have come up against the contradiction which Saccheri sought. To *prove* the contrary, Beltrami and Klein had recourse to the concept of a *model* of one theory in another: remember that the earliest of these "models" was the one which established the method of coordinates, by providing a "dictionary" which is used to "translate" any object or relation in Euclidean plane geometry into an object or relation formed from pairs (x, y) of real numbers, in such a way that any theorem in one of the two theories has

---

[9] It does not seem that Beltrami knew of Riemann's *Inaugural Lecture*.

a corresponding theorem in the other (Chapter III, 8). In this case, Beltrami and Klein constructed models of the plane non-Euclidean geometries of Bolyai and Lobachevski *in Euclidean geometry*. Slightly modified by Poincaré, the "dictionary" is as follows; with the "non-Euclidean" versions in the left-hand column and their Euclidean translations on the right:

Plane   Poincaré's half-plane H (Chapter V, 3, F)

Line   Semicircle centred on Ox or half-line parallel to Oy

Angle   Angle

Circle   Circle (contained in H).

The (non-Euclidean) length of a curve contained in H can be defined only by an analytic formula (Appendix 1), but it can be said that for a small segment of line in the neighbourhood of a point (x, y) of H, its non-Euclidean length is about equal to its Euclidean length divided by y, which means that the non-Euclidean lengths become *infinitely large* in the neighbourhood of Ox. In the same way, the non-Euclidean area of a small triangle in the neighbourhood of (x, y) is about equal to its Euclidean area divided by y^2. The geodesics for non-Euclidean lengths are precisely the semicircles centred on Ox and the half-lines parallel to Oy; on a semicircle, the non-Euclidean length of an arc $A_1 A_2$ is equal to

$$\log \tan \frac{\varphi_2}{2} - \log \tan \frac{\varphi_1}{2},$$

and on a half-line parallel to Oy, the non-Euclidean length of a segment $B_1 B_2$ is

$$\log y_2 - \log y_1,$$

as in Figure 51.

It is possible then to examine Euclid's axioms (or better, Hilbert's axioms) with the exception, of course, of the parallel postulate; we note that all the "translations" of these are *theorems* of Euclidean geometry (or, if we prefer, theorems of the theory of real numbers, by using coordinates). For example, the first "requirement" of Euclid is translated into the following theorem: through two distinct points A_1 and A_2 of H, not on a line parallel to Oy, there passes one and only one semicircle centred on Ox. This can be proved simply by taking the line Δ perpendicular to $A_1 A_2$ through its midpoint M. Since $A_1 A_2$ is not parallel to Oy, the line Δ is not parallel to Ox, and therefore meets it at a single point C, which is the centre of the semicircle we were looking for (Figure 51). The "indefinite" extension of the segment of semicircle $A_1 A_2$ is possible, and even Archimedes' axiom is verified on this semicircle, because the successive "multiples" of this segment have as their endpoints $A_2, A_3, ..., A_n, ...$, where

$$\log \tan \frac{\varphi_{n+1}}{2} - \log \tan \frac{\varphi_n}{2} = n \cdot \left(\log \tan \frac{\varphi_2}{2} - \log \tan \frac{\varphi_1}{2} \right).$$

Hence the angle φ_n tends to π when n increases indefinitely (but it should be noted that it *never* reaches the point A_∞ on the axis Ox; this point is *not* in H).

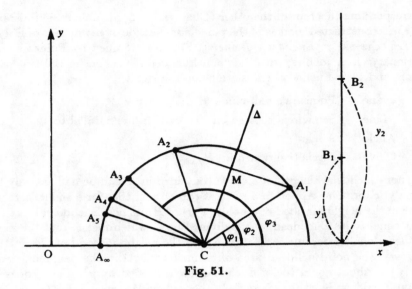

Fig. 51.

The determining factor in pursuing implications is the existence of the group of non-Euclidean "displacements": this is the group $GL^+(2, \mathbf{R})$ of which we spoke in Chapter V, 3, F, and which has the property of keeping unchanged the non-Euclidean lengths and angles (Appendix 1) and of being able to transform two points A and B of H into two other points A' and B' on the sole condition that the non-Euclidean distances $A'B'$ and AB are equal. Thanks to this group, it can be verified for example that the "cases of congruence of triangles" are valid in the non-Euclidean sense.

Once all the "non-Euclidean" axioms are translated into Euclidean theorems, we can rest assured that no contradiction will *ever* be deduced from these axioms, for in "translating" them, we would obtain a contradiction in Euclidean geometry; and, as Poincaré said, "nobody doubts that ordinary geometry is free from contradiction".

For the rest, it follows immediately that there are in H infinitely many lines "parallel" to a non-Euclidean line Δ, passing through a point B not on Δ: if, for example, B is outside a semicircle Δ, it is enough to consider all the semicircles with centre C' on Ox, passing through B and through any point of Ox not on the diameter PQ of Δ (Figure 52). A simple calculation shows too that for any non-Euclidean triangle ABC in H, the sum of the angles of the triangle is always less than two right angles, and in fact we have

$$\pi - (A + B + C) = \text{area}(ABC),$$

the area naturally being taken in the non-Euclidean sense (Appendix 1). It can, moreover, easily be seen that the sum $A + B + C$ can be as small as you like. It is also to be observed that in the non-Euclidean geometry of Bolyai-Lobachevski, there are no rectangles, because if Δ and Δ' are two non-

Euclidean "parallel" lines, there is *only one* non-Euclidean line which is at the same time perpendicular to Δ and to Δ' (Appendix 1).

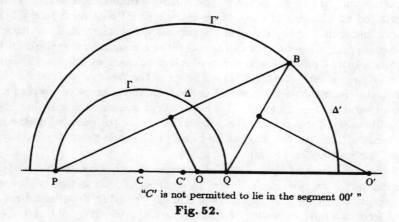

"C' is not permitted to lie in the segment OO'"

Fig. 52.

Klein showed that there is also a non-Euclidean geometry in which, as for the sphere, there are no "parallel lines", and the "length" of any line is bounded, but yet there is always *one single* geodesic passing through two distinct points (contrary to what happens with the sphere). But the surface on which such a geometry is defined cannot be embedded in ordinary space $\mathbf{R}^3$: it can only be embedded in $\mathbf{R}^4$.

While those who believed in "evident truth" had to accept the logical validity of non-Euclidean geometries, many still thought that it could be shown experimentally that only Euclidean geometry could provide a model of space in accord with physical reality. Poincaré devoted definitive pages (to which we refer) to making clear that this idea is fallacious: no experiment can prove the "truth" of a model based on a geometry, but only the concordance of this model with a certain physical theory: if the physical theory changes, then the geometric model must be changed to preserve this concordance [9].

2. The Deepening of the Concept of Number

A. Irrational Numbers

Up to the Renaissance, the predominance of geometry over the other areas of mathematics was uncontested[10]. We recall that Euclid felt the need to "represent" natural numbers as segments of a line, and that the problems which we reduce to quadratic equations were expounded by him in the form of plane

---

[10] A testimony of this pre-eminence is the name "geometer", which was given to *all* mathematicians up until the end of the nineteenth century, even to those who were not concerned with geometry!

geometric constructions to be worked out. The great expansion of algebra and analysis that took off in the seventeenth century and their applications to geometry, while "pure" geometry in the Euclidean manner remained a little stagnant, led to a change of view. "Geometric rigour" continued to excite admiration, and from the beginning of the nineteenth century, analysts, tired of the looseness and imprecision of their predecessors, sought to give the infinitesimal calculus the irrefutable character of a logical chain of inferences such as they found in geometry: this is what is known as the "return to rigour", which saw the establishment at the basis of analysis of the concept of *limit* of a sequence of numbers (see Chapter III, 9). In 1817, however, Bolzano, one of the standard-bearers of this reform, considered that "pure (or general) mathematics [consists of] arithmetic, algebra and analysis", while geometry is only an "applied (or special) part".

The basic concept of algebra and of analysis must be that of *number* (that is, our real numbers). Cauchy's *Course in Analysis* of 1821 in fact begins with this concept. It is true that he confines himself to saying that "numbers are derived from absolute measures of magnitude ... preceded by the signs + or −", and to enumerating operations on numbers and their properties, without any proofs; thus he remains at the same point as Descartes and Newton (who are tied directly to Book V of Euclid's *Elements*), apart from the fact that he states explicitly the "principle of nested intervals" (Chapter III, 6), which is indispensable to the "rigorous" proofs of the infinitesimal calculus. Since he gives no proof of this, it may be thought that he considered it to be obvious on account of its geometric interpretation; and Gauss must have thought the same when he stated without proof an equivalent theorem, the existence of a least upper bound of a bounded sequence. But Bolzano, who himself likewise stated a form of this same principle in 1817 (before Cauchy), considered it "an intolerable fault against good method" to have recourse to "the evidence of a geometric theorem" to prove a property of numbers. All the same, he does not produce any other method of proving the existence of a number common to a sequence of nested intervals, and has to confine himself to saying that the affirmation of this existence "has nothing impossible in it"!

Where in fact, other than in geometry, can one find a proof of the "existence" of irrational numbers? The Greeks were well able to define relations between these numbers (equality, inequality, addition: see Chapter III, Appendix 1), but they *accepted* their existence. From 1820 the idea began to emerge that what should be put at the basis of classical mathematics is not the geometrical concepts of the Greeks but the concept of *natural number*. This was the idea expressed by M. Ohm as early as 1822, at the beginning of an ambitious *Treatise*, which undertook to write for the whole of mathematics the equivalent of Euclid's *Elements*. Here he reproaches Cauchy for not having constructed a theory of real numbers on the basis of the natural numbers, but he himself was quite incapable of building one. According to Dedekind's report, Dirichlet affirmed that any theorem in algebra "can be stated as a theorem about the natural numbers". But it was not until about 1860 that "constructions" of real numbers began to appear. From our point of view, these can be considered as *models* of the axiomatic theory of real numbers (see Chapter III, Appendix II) in the theory of natural numbers, just as we have seen *models* of geometries (Euclidean or non-Euclidean) being constructed in the theory of real numbers (see Chapter III, 8, and, above, 1, C). In this way *all* of classical mathematics had models in the theory of natural numbers: this was called the *arithmetization* of mathematics, which was so successful at the end of the nineteenth century that it was even taught to students before they entered university, which meant that they had to study arguments which were of hardly any use to them afterwards. We shall confine ourselves to describing summarily, in Appendix 2, the principles of two of these constructions, those of Dedekind and of Méray-Cantor.

B. Monsters

Although it cannot be affirmed that there was a relation of cause and effect, the construction of these models must have reassured analysts of the second half of the nineteenth century, confronted as they were at that very time with what can be called the arrival of "monsters", mathematical phenomena that were completely unforeseen and altogether antithetical to what our "intuition" of the nature of space teaches us to expect.

The earliest chronologically of these is concerned with tangents to curves. Up until the nineteenth century, the concept of a "plane curve" remained imprecise. After the introduction of coordinates, the curves which were considered were defined by a relation $F(x, y) = 0$ between the coordinates, or by a "parametric representation"

$$x = f(t), \quad y = g(t)$$

suggested by the trajectories of a mobile point varying as a function of "time" t. But the functions F, f and g had to have imprecisely specified properties of "regularity", for example, that of being combinations of "elementary" functions. In any case, it seemed to go without saying that it is possible to draw a plane curve by the movement of a pencil on paper, which implies a "direction" of this movement at every point. Mathematically, this is expressed as the existence of a *tangent* to the curve at every point, with the possible exception of a finite number of points, which were termed "angular".

At the beginning of the nineteenth century, once the concept of a *continuous* function had been precisely formulated, the question took the following form: has a continuous function $y = f(x)$ defined in an interval of **R** a *derivative* (Chapter III, 9) at every point of the interval, with the possible exception of a finite number of points? Many people accepted that it has, without discussion. Cauchy did not pronounce on the question, and confined himself for the most part to considering only functions with a derivative that was assumed to exist. There were, however, some attempts at proof, among others by Ampère and even by Galois when he was still a student at the Ecole Normale. Then, in a manuscript written about 1830, but not published until a century later,

Bolzano gave the definition of a continuous function in an interval $[a, b]$ which can be proved to have *no* derivative at *any point* [11]. It was not until 1875 that another example was published of a continuous function with no derivative at any point: this was the one described by Weierstrass in his lectures beginning in 1872. Later, Banach was able to show that in an entirely reasonable sense, there are "many more" such functions than there are differentiable ones (Chapter V, 4, C). This was only a beginning, and "counter-examples" of this kind became so numerous that it has been possible to write whole volumes on them. We will confine ourselves to citing, among the more bizarre of these: a *differentiable* function, not constant in any interval, which nevertheless has an infinite number of maxima and minima in every interval; and a "surface", in space, such that it is a "party wall" between *three* "rooms" at the same time, which is somewhat hard to imagine! But the strangest is undoubtedly the "curve" invented by Peano: it is defined by two *continuous* functions

$$x = f(t), \quad y = g(t) \quad \text{for} \quad 0 \le t \le 1$$

and *fills the square* $0 \le x \le 1$, $0 \le y \le 1$ (in other words, for *any* point (a, b) of this square, there is at least one value t_0 of t such that $a = f(t_0)$ and $b = g(t_0)$).

What do these examples show? Simply, in my opinion, that any connections between our supposed "intuitive understanding" of space and the axioms which define the mathematical objects of geometry are entirely superficial, contrary to what was thought by the mathematicians of the seventeenth and eighteenth centuries. To speak of the "truth" of these axioms, in the sense in which the word is commonly understood, therefore seems absurd.

C. The Axiomatization of Arithmetic

It is rather curious that Euclid did not think he needed to give an axiomatic presentation of his Books of arithmetic in the *Elements*: did he perhaps believe that it was useless to introduce "hypotheses" for basic notions accepted without question? In the Middle Ages, Campanus (thirteenth century) sought to fill this gap, but his system of axioms is incomplete. Up until about the middle of the nineteenth century, books of arithmetic merely followed Euclid. The first innovator in this area was H. Grassmann, in 1861. He noted that, to define addition and multiplication in the set $\mathbf{N}^*$ of integers > 0 (the "natural numbers"), it was enough to use the operation $n \mapsto n + 1$; in fact, if $n + m$ is defined, we set, by definition, that

$$n + (m + 1) = (n + m) + 1.$$

For multiplication, in the same way we take $n \cdot 1 = n$, and, if $n \cdot m$ is defined, then we take

---

[11] Bolzano said only that in any interval, no matter how small, there are points where the function has no derivative.

$$n \cdot (m+1) = n \cdot m + n.$$

From this, Grassmann proved the associativity and the commutativity of addition, the associativity and the commutativity of multiplication, and its distributivity with respect to addition. For example, to prove that $a + (b + c) = (a + b) + c$, he notes that it is true for $c = 1$, by virtue of the definition of addition; and, if $c = x + 1$ and if we deem it proved that $a + (b + x) = (a + b) + x$, then we can write

$$a + (b + c) = a + (b + (x + 1)) = a + ((b + x) + 1)$$
$$= (a + (b + x)) + 1 = ((a + b) + x) + 1$$
$$= (a + b) + (x + 1) = (a + b) + c.$$

It will be observed that Grassmann uses implicitly what is known as the *principle of recurrence* (or "principle of complete induction" as it was called in the nineteenth century), just as mathematicians had been doing since Pascal's time, without stating it in its general form: when we wish to prove a property $P(n)$ for all natural numbers n, it is enough to prove $P(1)$, and to prove that the property $P(n)$ *implies* $P(n + 1)$.

When Dedekind published, in 1872, his little work on the definition of irrational numbers (Appendix 2), he did not yet feel the need of an axiomatic presentation of natural numbers. It was not until a little later that he thought of it, in connection with his study of the concept of infinite set, which we shall look at in more detail (§3). There he established a system of properties of natural numbers, which Peano, in 1889, reduced to a set of axioms, which are now known as the *Peano's axioms*. His presentation is the same as that of Pasch and Hilbert for geometry (Chapter III, 4): the basic objects are a set $\mathbf{N}^*$, an element $1 \in \mathbf{N}^*$, and a mapping $s : \mathbf{N}^* \to \mathbf{N}^*$, such that the following three axioms hold:

I) s is *injective*, in other words, if $s(a) = s(b)$, then $a = b$;

II) $s(a) \neq 1$ for each a;

III) if E is a subset of the set $\mathbf{N}^*$ such that $1 \in E$ and $s(E) \subset E$, then $E = \mathbf{N}^*$.

Of course, $s(n)$ is what will be written $n + 1$ once addition is defined: Axiom III) is equivalent to the principle of recurrence (taking for E the set of $n \in \mathbf{N}^*$ such that $P(n)$ is satisfied). Taking Dedekind as his model, Peano showed that his axioms imply all the properties of Euclid and Grassmann for the members of $\mathbf{N}^*$, as do also two results which can be said to justify the unique place accorded by mathematicians to natural numbers and to their properties:

1) If $(\mathbf{N}'^*, 1', s')$ is a second system of basic objects satisfying Axioms I), II), and III), then there exists *one and only one bijection* $f : \mathbf{N}^* \mapsto \mathbf{N}'^*$ such that

$$f(1) = 1' \quad \text{and} \quad s'(f(n)) = f(s(n)).$$

If Peano's axioms are seen as defining a *structure* (Chapter V, 3), it can be said that, if two structures satisfy these axioms, then there exists a unique isomorphism of the one onto the other; or, to be less precise, there is *only one* set of natural numbers.

2) None of the three axioms is the consequence of the other two: this is proved by producing three basic objects which satisfy two of the axioms but not the third[12].

We note an important property of natural numbers, following from Peano's axioms: for every non-empty set A of natural numbers, there exists a *smallest* member of A for the usual order of natural numbers[13].

3. Infinite Sets

A. Infinite Sets and Natural Numbers

Etymologically speaking, "infinite" is the negation of "finite"[14]. According to intuition, a finite set is a set of objects which can be *counted*: mathematically speaking, once the theory of natural numbers had been set up on the basis of Peano's axioms (§2, C), a finite set F became a set such that there exists a *bijection* of F onto one of the subsets Z_n of N^* formed of the numbers m such that $1 \leq m \leq n$. The theory of natural numbers shows that there cannot exist any bijections of Z_p onto Z_q for $p \neq q$, and thus the number n is uniquely determined: it is the "number of members", or "cardinal number" of F. *A fortiori*, there can be no bijection of N^* onto a set Z_n: in other words, in the theory of natural numbers, N^* is an *infinite* set.

It is possible to construct a theory of finite sets without using Peano's axioms: we need only say that F is finite if there does not exist a bijection of F onto the complementary set $F \setminus \{a\}$ of any member $a \in F$. Euclid's arithmetic can then be developed using this definition and the operations of the "naïve" theory of sets: union, complementary set, and product (Chapter V, 3, B), without mentioning infinite sets.

By contrast, Euclidean geometry necessarily implies that the set of points of a line is infinite, because "between" two points there is always another one, and because an operation consisting of for instance taking the mid-point of a segment can be repeated indefinitely. Since Aristotle did not accept the

---

[12] For an example which satisfies I) and II) but not III), all we need to do is take the set N^* itself as the basic set, but set $s(n) = n + 2$. For an example which satisfies I) and III), but not II), we take the set $E = \{1, a, b\}$ formed of three members as the basic set, and $s(1) = a$, $s(a) = b$, and $s(b) = 1$. Finally, for an example satisying II) and III), but not I), we take the same set E, with $s(1) = a$, $s(a) = b$, and $s(b) = a$.

[13] We need only consider the case where A is finite, because if $a \in A$, we need only show that the set A' of natural numbers $n \in A$ such that $n \leq a$ has a smallest member. We can then argue by induction on the number of members of A'.

[14] We are not thinking here of the sense of "unlimited" which the word "infinite" sometimes bears.

existence of infinite sets (or what he calls "actual infinity"[15]), he found a way out by saying that a line "is not composed" of points, or in other words, it is not possible to speak of the set of points of a line! For the sake of Aristotle's good name, it is to be hoped that the reason which is given for this ukase is only the work of a disciple, because it is a fine example of mental confusion: a point, according to this "argument" (Phys. VI, 1, $231^a21 - 232^a22$) should be "in contact" with those which are on the half-line bounded by the point, and this is impossible, since the "contact" could only be with a "part" of the point, and a point "has no parts"!! The author has invented out of nothing this "contact", which does not appear in any axiom, and has confused a line with a real object.

The first more serious argument claiming to justify the taboo on "actual infinity" is probably that to be found in the work of Galileo: if we pair each natural number n with its square n^2, then we define (in our language) a bijection of N^* onto a proper *subset* of N^*; but, says Galileo, if there were "as many" squares as natural numbers, that would violate the axiom that "the whole is greater than the part"; therefore we cannot say that the natural numbers form a set! It is surprising to see Cauchy still citing this argument with approval. But, in 1859, there appeared the posthumous work of Bolzano, *Paradoxes of the Infinite*. This work is chiefly designed to expound the author's philosophical views; but is is here that we find, before Dedekind and Cantor, a clear distinction made between inclusion and equipotence: there is, he says, a bijection of the interval $[0,5]$ of the line R onto the interval $[0,12]$, namely, the mapping $x \mapsto 12x/5$, even though the interval $[0,5]$ is a proper subset of $[0,12]$.

The first systematic studies of infinite sets are due to Dedekind and Cantor, and were carried out at much the same time, beginning about 1872, even though Dedekind's work expounding his findings was not published until 1888. As we have said (see Chapter V, 3, B), it is here that he codifies the basic operations of "naïve" set theory, with the aim of clarifying the relations between the concept of an infinite set and the concept of a natural number, which are the main subject of his book. The mathematical *definition* of an infinite set, which he gives for the first time, is precisely what appears in the examples of Galileo and of Bolzano: a set E is called *infinite* if there exists a bijection of E onto a subset A of E *distinct* from E. Dedekind then proved, *without assuming* the existence of the natural numbers, that *any* infinite set E necessarily contains a subset N^*, which has a member $1 \in N^*$ and for which there is a mapping $s : N^* \to N^*$ *satisfying Peano's axioms* (stated for the first time by Dedekind, as we have said). The principle of his ingenious proof is the following: by hypothesis, there exists an element $a \in E$ and a bijection f of E onto a subset $f(E)$ of E *not containing* a. Dedekind used the term *chain* (for f) for a subset K of E containing a and such that $f(K) \subset K$

---

[15] Aristotle (and many mathematicians after him) accepted only "potential infinity", which means that for numbers with a certain property, when a finite number of these has been obtained, there are always more. "Actual infinity" would be the totality of these objects.

(since $f(K) \subset f(E)$, $f(K)$ does not contain a). The set N^* is simply *the intersection of all the chains*, and if we write 1 instead of a, the restriction s of f to N^* and the element 1 satisfy Peano's axioms. We shall not give the details of Dedekind's proof, nor of the way in which he obtains all the properties of natural numbers by arguments from pure set theory.

B. The Comparison of Infinite Sets

The contributions of Cantor to set theory are even more original than those of Dedekind. Before he did his work, only two kinds of sets were recognized: finite sets and infinite sets, and no attempt was made to differentiate among the latter. It was Cantor who, to general surprise, was the first to prove that there are *several* kinds of infinite set, which cannot be reduced to one another.

We have described (see Chapter V, 4, A) how Cantor was led to introduce the relationship of *equipotence* between sets. In particular, he distinguished in the first place, among infinite sets, those which are *countable*, defined as equipotent to the set N^* of natural numbers. Their importance comes from the part played in analysis by *sequences* of distinct members of a set (if (a_n) is a sequence, the mapping $n \mapsto a_n$ can be considered as a bijection of N^* onto the set of members of the sequence). Cantor saw very soon that the set of *pairs* (m, n) of natural numbers is countable: his procedure is easily "visualized" if we "count" the pairs following the "rows parallel to the second bisecting line" (Figure 53).

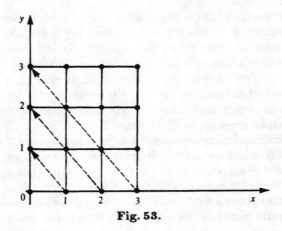

Fig. 53.

It is more convenient here to consider instead of N^* the set $\mathsf{N} = \mathsf{N}^* \cup \{0\}$, so that Cantor's "counting" gives the sequence

$$(0,0), \quad (1,0), \quad (0,1), \quad (2,0), \quad (1,1), \quad (0,2),$$
$$(3,0), \quad (2,1), \quad (1,2), \quad (0,3), \quad \ldots$$

An immediate consequence, although already rather a surprising one, is that the set $\mathbf{Q}^*_+$ of irreducible *fractions* p/q is countable[16]: all we need do in fact is list the pairs (p,q) following Cantor's procedure, keeping only those which are not zero, and which have no common factor other than 1.

In 1873, Cantor put to himself a problem which no-one else had thought of: is the set $\mathbf{R}$ of real numbers countable? His first major discovery was that the answer to this question is *negative*. He found this out by an argument by contradiction as simple as it is ingenious, based on the principle of nested intervals (Appendix 3).

This success encouraged Cantor to devote a large part of his career to problems of equipotence. On the model of the "cardinal numbers" of finite sets, he said that two sets have the "same power", or the "same cardinality" if they are equipotent, and he set himself to define a relation of *order* among cardinalities. The natural inclination is to say that a set E has a cardinality "less than or equal to" that of a set F if there exists a bijection of E onto a *subset* of F. For finite cardinalities (that is to say, the natural numbers), this is indeed the usual relation $m \leq n$; and the cardinality of every finite set is indeed less than (and not equal to) the cardinality of the countable sets, which is often denoted by the Hebrew letter $\aleph_0$ (aleph zero). But before we can speak of a relation of total order, we must have verified the four axioms which characterize these relations (Chapter V, 3, D). Reflexivity is clear, and transitivity is immediate: if there is a bijection f of E onto a subset of F and a bijection g of F onto a subset of G, then the composed mapping $g \circ f$ is a bijection of E onto a subset of G. Cantor however ran into difficulties with the other two axioms. Not until 1898 did Schröder and F. Bernstein prove that, if two sets E and F are such that there is a bijection of E onto a subset of F, and likewise a bijection of F onto a subset of E, then E and F are equipotent[17] (Appendix 3). As for the last axiom, the fact that for any two sets E and F, there is always *either* a bijection of E onto a subset of F *or* a bijection of F onto a subset of E, this can be proved only by introducing into set theory an axiom of which we shall speak later (§3, B), known as Zermelo's axiom.

Cantor and his successors also developed an "arithmetic" of infinite cardinalities, which has obvious differences from ordinary arithmetic. The task is to calculate the cardinality of a set obtained, beginning with sets of given cardinality, by the usual processes of set theory: the union of sets without any common members, product, and the set of subsets. If m and n are the cardinalities of E and F, respectively, then we denote by $m + n$, $m \cdot n$, and 2^m the cardinalities of the sets $E \cup F$, $E \times F$, and $\mathfrak{P}(E)$, respectively; when m and n are *infinite* cardinalities, we have

(1)
$$m + n = m \cdot n = \sup(m, n),$$

where the notation $\sup(m, n)$ designates the larger of the two cardinal numbers m and n. Further, for any cardinality m, we have

---

[16] The fact that between any two rational numbers there are always infinitely many other rationals would seem to stand in the way of their being "enumerated"

[17] A proof of this theorem dating from 1887 has been found among Dedekind's papers.

(2) $$\mathfrak{m} < 2^{\mathfrak{m}}.$$

This latter relation shows that, if we consider successively the sets N, $\mathfrak{P}(\mathsf{N})$, $\mathfrak{P}(\mathfrak{P}(\mathsf{N}))$, $\mathfrak{P}(\mathfrak{P}(\mathfrak{P}(\mathsf{N})))$, ..., then we obtain infinite sets whose cardinalities are *all different*. But the relations (1) show that it would be pointless to speculate on a true "arithmetic of the infinite" on the model of ordinary arithmetic.

The special case $\mathfrak{m}^2 = \mathfrak{m}$ of (1) for an infinite cardinality $\mathfrak{m}$ gives us again Cantor's initial result on $\mathsf{N} \times \mathsf{N}$ when $\mathfrak{m} = \aleph_0$. Before this result was proved for an arbitrary infinite cardinality, Cantor in 1877 proved it for the cardinality c of the set R of real numbers: the sets R and R^2 are *equipotent*. This was the theorem of Cantor's which caused most surprise and even disarray among contemporary mathematicians. Ever since the Greeks, the conviction had held of the profound difference between geometric objects of one, two or three "dimensions" (curves, surfaces and volumes): Cantor's result seemed to abolish these distinctions, and hence to destroy the whole of geometry! It was Dedekind who saw how to reconcile Cantor's theorem with classical mathematics. The concept of dimension has to do not only with the cardinality of R^n, but also with its *topology* (Chapter V, 3, E); the bijections $\mathsf{R} \mapsto \mathsf{R}^n$ defined by Cantor are not continuous, and Dedekind conjectured that, if $m \neq n$, there could not exist any bijection $f : \mathsf{R}^m \to \mathsf{R}^n$ which is a *homeomorphism* (Chapter V, 4A). But the proof of this conjecture for all pairs (m, n) such that $m \neq n$ was not discovered until 1911 by L. E. J. Brouwer, using completely new ideas introduced by Poincaré and by himself.

It is possible to define in a general way, for many structures, a concept of *dimension* (or even several such concepts), that is to say, a real number attached to the structure and *invariant* with respect to isomorphism (Chapter V, 4, B). This is the case not only in topology and in homological algebra (Chapter V, 5, A, VIII, IX and X) — where there can be several different definitions which do not give the same number — but also in algebraic geometry (*ibid.*, XX) and in measure theory (*ibid.*, XII). In this last case, the "dimensions" can take values which are not natural numbers (and may even be irrational), and they have recently acquired importance in certain applications, with what are known as "fractal" structures. When a set is equipped with several structures, there is no longer any reason for the "dimensions" attached to these structures to coincide.

4. "Paradoxes" and their Consequences

A. Existence and Constructions

The words "exists" and "does not exist" are among those most often used by mathematicians. In view of the nature of mathematical objects, it is clear that when we say that a mathematical object with a certain property P "exists", we do not mean the same thing as when we say that something in the world of experience "exists". In the latter case, the proof of "existence" consists in doing an experiment which makes the object in question accessible to our senses, or enables us to infer this "existence" by logical consequences accessible to our senses — which means in this case that a theory intervenes between the object and our senses.

From the point of view of the axiomatic theory introduced by Pasch and Hilbert — which is the one adopted by the huge majority of mathematicians of our day — , the concept of "existence" is relative to a system of axioms. It is a question of the properties P formulated with the help of the basic relations of the theory, and dealing with objects which are basic or defined within the theory: the set E of objects which have the property P is then defined within the theory, and to say that "there exists" an object with the property P means that the statement "E is not empty" is a theorem within the theory, following from the axioms[18]. We have just seen (§3, B) two examples in the theory of real numbers: the set of bijections of $\mathbf{R}^m$ onto $\mathbf{R}^n$ for $m \neq n$ is not empty (Cantor), but the set of homeomorphisms of $\mathbf{R}^m$ onto $\mathbf{R}^n$ is empty (Brouwer).

To prove that E is empty, we usually argue by contradiction, showing that the contrary hypothesis leads to a contradiction in the theory. Brouwer proceeds in this way in the preceding example. We recall that the first known argument "by contradiction" (see Chapter III, 2) proves precisely that the set of *rational* numbers r such that $r^2 = 2$ is empty.

By contrast, the existence of an object with a certain given property is often proved by means of a "construction". This is the case for Cantor's theorem on the equipotence of $\mathbf{R}$ and $\mathbf{R}^2$ (Appendix 3). A much simpler example is the existence of rectangles in Euclidean plane geometry, which we have already described (§1, A).

There are nevertheless proofs of "existence" which are done by means of contradiction. One of the most recent and most remarkable is that of the theorem of Feit and Thomson (see Chapter V, Appendix 2, D). The problem is to prove that any finite group G of *odd* order is solvable. This is equivalent to saying *(loc. cit.)* that there exists in G a distinguished subgroup which is different from G and from $\{e\}$. A "constructive" proof would consist in "constructing" such a subgroup, beginning from the fact that the order of G is odd; but no way of doing this is known at present. Thus the argument is by contradiction: suppose that there exists a group G_0 of odd order $\neq 1$ which has *no* distinguished subgroup other than itself and $\{e\}$ (in other words, a *simple* group); we can even suppose that the order of such a group G_0 is the smallest possible. The proof then consists in analysing the properties of G_0 until, after 250 pages, a contradiction is reached.

Most mathematicians prefer "constructive" proofs of existence, which often give more precise information about the objects "constructed"; but they accept "non-constructive" proofs when there are no others.

---

[18] The convention is that when a structure has sets among its basic objects, these are not empty.

B. The Mixed Fortunes of the Concept of Set and the Axiom of Choice

Until about 1870, when mathematicians spoke of sets of objects it was always with reference to *mathematical* objects, especially "classical" objects, such as numbers, "figures", functions, etc. Since about 1930, this is again what the vast majority of mathematicians do. A curious episode, however, was to take place in the interval, in which set theory became the meeting-place of mathematics and philosophy, and which was to have the effect of disturbing the minds of many mathematicians. This is the alleged "crisis of foundations".

To begin with, Cantor and Dedekind must themselves no doubt bear the blame, as it was they who felt the need to "define" a set. We know Cantor's "definition": "a grouping together of distinct objects of our intuition or of our thought"; and Dedekind's is hardly any different. They are just as empty and unusable as Euclid's pseudo-definitions of the point and the line (see Chapter III, 4). But they open the door to a way of thinking according to which the word "set" does not denote only a mathematical object E, which, together with other mathematical objects x, has the relation of belonging $x \in E$; it can also bear its original meaning of a collection of material objects, such as words or sentences in a written text.

In arguing about these sets, every mathematician of this period let himself be guided by his peculiar "intuition" about them, in the same way as the mathematicians of the seventeenth century argued about the "infinitely small". But apart from a certain blurring in the way in which Cantor and his successors defined the cardinal numbers and some relations of order, these arguments depended only on the "naïve" language of sets (Chapter V, 3, B) and did not give rise to controversy.

Nevertheless, towards the end of the century, it was realized that the operations on sets allowed by Boole and Dedekind are not completely sufficient as a basis for many arguments[1] in analysis. For example, let E be a set of points in the plane, and x_0 a point in the plane not belonging to E but such that any disc with centre x_0 contains at least one point of E. The problem is to prove that there exists a *sequence* $a_1, a_2, ..., a_n, ...$ of points of E which has as its *limit* x_0. The argument is very simple: let D_n be the disc with centre x_0 and radius $1/n$ (Figure 54). There is a point $a_1 \in E$ in the disc D_1, a point $a_2 \in E$ in the disc D_2, etc. These points form the required sequence, since for any natural number N, the distance from x_0 to a_n is $\leq 1/N$ whenever $n \geq N$. If we generalize from the properties peculiar to this example, we see that in fact the following general principle is being applied. Let (E_n) be a sequence of subsets of a set E ($E_n = E \cap D_n$ in the preceding example), which *are not empty*: there is then a mapping $n \mapsto a_n$ of $\mathbf{N}^*$ in E such that $a_n \in E_n$ for any n. This may seem an altogether natural conclusion, and the principle was for a long time applied without even being mentioned. The first to draw attention to it was no doubt Peano in 1890, when he had to apply it to a set E of functions. The point which he raised is that, for any n, the member a_n is by no means uniquely determined in E_n (E_n will often be an infinite set). If one argues "intuitively", it is thus necessary to "choose" a_n in E_n, and therefore to make *infinitely many* successive "choices", which Peano refused to do. He was to be followed in this by others at the beginning of the twentieth century, especially after Zermelo had stated, under the unfortunate name of the "Axiom of Choice", a far more general and less "intuitive" principle. In view of applications to the theory of cardinalities, he needed the following statement: given a mapping $U : I \to \mathfrak{P}(E)$ of an *arbitrary* set I (not necessarily countable) into the set of subsets of a set E, such that $U(\alpha)$ is *not empty* for each $\alpha \in I$, there exists a mapping $f : I \to E$ such that $f(\alpha) \in U(\alpha)$ for each $\alpha \in I$. This mapping thus "chooses" a member of $U(\alpha) \subset E$ for each $\alpha \in I$. The case where I is countable is the principle rejected by Peano, sometimes called the *Axiom of Countable Choice*.

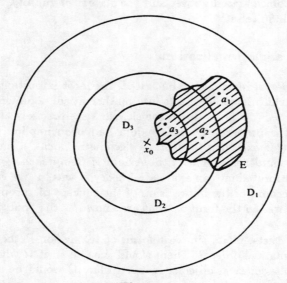

Fig. 54.

It can be seen where this tendency was leading: the tendency, that is, to attribute a character of "reality" to the concept of set, forgetting its nature as a mathematical object. A set is not a display in a shop from which one "chooses" objects, any more than a line is a thread stretched between two points; and it is no more reasonable to reject Zermelo's axiom — which does no more than affirm the existence of a mathematical object, a mapping — than it would be to refuse to accept Euclid's first "requirement", on the grounds that it is not possible to stretch a thread between the earth and Sirius!

So it was that Zermelo thought the best way to escape from futile controversies would be to restore to the concept of set its character as a mathematical object (in Plato's sense), by defining it by means of a system of axioms, as Pasch and Hilbert had done for geometrical objects (Chapter III, 4). There is only one "basic" relation $x \in X$ between any mathematical objects. From this relation is derived another, the relation of inclusion $X \subset Y$, which means that any object x which satisfies $x \in X$ likewise satisfies $x \in Y$. Zermelo then takes the operations of Cantor's and Dedekind's "naïve" language of sets, and states in the form of axioms the properties which they used. First the Axiom of *Extension* already stated by Boole and Dedekind, namely, the fact that the relation "$X \subset Y$ and $Y \subset X$" implies that $X = Y$. Then we have a series of axioms affirming the *existence* of the empty set $\emptyset$ (such that $x \notin \emptyset$ for any object x); of the set of the pair (a set $\{x, y\}$ whose only two members are two arbitrary objects x and y (distinct or not)); of the set $\mathfrak{P}(X)$ of subsets of a set X (a set whose members are all the subsets $U \subset X$, so that the relation $U \in \mathfrak{P}(X)$ is equivalent to $U \subset X$). He adds to these the Axiom of Choice stated above, and the *Axiom of Infinity*, affirming the existence of an infinite set[19].

C. Paradoxes and Formalization

One axiom, however, remained to be stated, the most important of all, since it was this one which introduced sets into mathematical arguments: it affirms that *any property* P defines a set whose elements are precisely all the objects which possess the property P. This axiom had been implicitly used since classical times (for example, to define a "geometric locus"), and was made explicit by Frege under the name of the Axiom of *Comprehension*. But before it could be incorporated into an axiomatic theory, it was to give rise to serious difficulties on account of the imprecision of the concept of "property". These difficulties appeared in the form of what are known as the "paradoxes" of set theory.

The first consists, in its simplest form, of taking for P the property of "being a mathematical object". There would thus be a set Ω whose members would be *all* mathematical objects. In particular, Ω would be a member of itself, which is somewhat hard to imagine. But worse is to come: every *subset* U of Ω is a mathematical object, and therefore a *member* of Ω; the set $\mathfrak{P}(\Omega)$ of these subsets would be a *subset* of Ω; but this contradicts Cantor's theorem which affirms that there is no bijection of $\mathfrak{P}(\Omega)$ onto a subset of Ω (§3, B, Formula (2))[20].

To escape this contradiction, Zermelo imported a restriction into his formulation of the Axiom of Comprehension: he does not accept the existence of a set formed from objects with a given property P unless these objects are *already* members of a set previously defined. For example, it is possible to speak of the set of even numbers, because we are talking about objects in a set already defined, the set of natural numbers, and about the property of "being divisible by two" for an element of this set.

---

[19] Bolzano and Dedekind had thought that it was possible to *prove* the existence of an infinite set, by considering for example "the set of thoughts"!

[20] This argument is a simplified form of the "paradox" known as "Russell's paradox".

But how was the word "property" to be understood at the level of generality at which Zermelo was working? Zermelo confined himself to saying that "whether or not a property is valid must be determined in a non-arbitrary way, by the axioms and by the universally valid laws of logic". Obviously he had in mind the types of property considered by mathematicians up till then. But a second "paradox" showed that this was not precise enough. We work in the set $\mathbf{N}^*$ of natural numbers, and a "property" of natural numbers is defined in the following manner. Consider all the sequences of words in the English language which constitute phrases unambiguously defining a natural number: for example, "the largest prime number dividing two hundred and forty-two". There is a finite number of English phrases with at most twenty words, therefore the same is true of those which define a natural number. The numbers *not defined* by one of these phrases are those which satisfy the property P: "not capable of being defined in a phrase of less than twenty words".

These numbers ought to form a non-empty set of natural numbers, which therefore has a smallest member n. Now this number is defined unambiguously by the phrase: "the smallest natural number which cannot be defined in a phrase of less than twenty words"; but this phrase has only sixteen words, and thus we have come up against a contradiction.

Although this argument looks strangely like a riddle, and although "properties" such as P had never before been thought of by mathematicians, it was necessary to formulate limitations which would correspond to the usage of mathematicians, and would prevent the statement of parasitical "properties". The solution proposed by Fraenkel and Skolem in 1922 consists in *eliminating* everyday language in the statement of mathematical relations, replacing it by an artificial language, formed from combinations of signs taken from a fixed list, and subject to an inflexible syntax which would avoid the ambiguities of ordinary speech. Such languages had been described in the nineteenth century by the founders of mathematical logic (§5, A), and are known as *formal* languages. Nobody has yet formulated in this language a property which would result in a "paradox".

Nevertheless, the mathematical arguments expressed in this language would be prohibitively lengthy. Numerous abbreviations have had to be introduced to arrive at the language in which current mathematical works are written. Experienced mathematicians know how to recognize texts which could be translated into formal language: where there are doubts, it is up to the author to bring in as much precision as is necessary to convince his colleagues.

Zermelo's axioms thus made precise and written in formal language constitute what is often called the ZFC system. The notation ZF is also used for this system without the axiom of choice AC. The vast majority of mathematicians nowadays use the system ZF or ZFC, usually without mentioning it explicitly. We have seen that, as far as these mathematicians are concerned, all branches of mathematics can be attached to set theory, and hence, in the last analysis, to the system ZF or ZFC. Such mathematicians are often called "formalists", even though they practically never use a truly formal language in their writings. They formulate them in the "naïve" language of Dedekind and

Cantor, without being under any illusion as to the falsely "intuitive" nature of the axioms, as soon as they come to deal with infinite sets. We shall have a few words to say (§5, D) about "non-formalist" mathematicians.

5. The Rise of Mathematical Logic

A. The Formalization of Logic

The use of logic is obviously not limited to mathematics, and it is displayed in all the texts of the Greek philosophers which have come down to us. We are indebted to Aristotle for having begun to systematize and codify the rules, and for having recognized that these rules are independent of the particular nature of the objects or the relations with which they deal. He introduced the *negation* of a proposition or of a relation, but his main preoccupation was the classification of syllogisms, which is not enough to cover all uses of logic, notably in mathematics. Two other logical operations have a large part to play in it, the *conjunction* "R and S" of two relations R and S, and their *disjunction* "R or S" [21].

With the development of algebra, the analogy between the rules of logic and those of algebraic calculation could not fail to be noted, and numerous attempts were made to create a symbolic script which would represent logical operations: those of Leibniz are particularly noteworthy, but they were not published until the twentieth century. Not before Boole and his immediate successors, in the middle of the nineteenth century, did anyone succeed in forging a calculus which is the precursor of the language of sets. Boole associates with a property of the objects of a "universe" the "class" of objects of this universe which possess the property. If x and y are the classes corresponding to the properties P and Q, the class which corresponds to the property "P and Q" is denoted by xy (intersection) and that which corresponds to "P or Q" is denoted by $x + y$ [22](union). We also need a notation x' for the class corresponding to the negation "not P" (the complement of x). Finally, we denote by 1 the class of all the objects of the universe, and by 0 the empty class, complementary to 1.

This constitutes what is known as "Boolean algebra", in which identities between the operations introduced are satisfied. Some of these, such as

$$x + y = y + x,$$
$$xy = yx,$$
$$x(y + z) = xy + xz,$$

---

[21] Despite this latter name, to say that "R or S" is true means that R is true, or that S is true, or that *both* are true, contrary to the "disjunctive" sense that the word "or" often has in current speech.

[22] Initially, Boole used the sign + only when the intersection xy is empty; Jevons generalized its use to cover all cases.

resemble those of ordinary algebra, but others, such as

$$x + x = x\,,$$
$$xx = x\,,$$
$$x + yz = (x + y)(x + z)\,,$$
$$1 + x = 1$$

are more unexpected. There are also the equations of "duality":

$$(x')' = x\,, \quad (x + y)' = x'y'\,, \quad (xy)' = x' + y'\,.$$

All these equations can moreover be deduced from a small number of them, and it may be considered that, by taking a small number of the latter as axioms, an algebraic structure has been defined (see Chapter V, 5, C).

All the same, Boolean algebra is not well adapted to the translation of mathematical theorems into formal language, and notably so when the theorems deal with relations between several objects, as for example in the following assertion:

For any positive real number x, there exists a real number y such that $x = y^2$.

In a formal language very often used by logicians of our day, essentially stemming from Frege and Peano, this is translated as

$$\forall x \ (x > 0 \quad \Rightarrow \quad \exists y \ (y \in \mathbf{R} \ \text{ and } \ x = y^2))\,.$$

The logical signs $\forall$ and $\exists$, called *quantifiers*, replace the words "for any" and "there exists", respectively, and $\Rightarrow$ means "implies that". Experience proves that it is possible to write all mathematical propositions coming under the rules of the ZFC system (§4, C) in this way. If no abbreviations were introduced (such as $>$, 0, $\mathbf{R}$, and y^2 in the above example), all that would be needed would be the logical signs and the mathematical signs $=$ and $\in$.

B. Metamathematics

The appearance, in the course of the nineteenth century, of varied structures without any link with experiential reality (see Chapter V, 3), had made obvious the freedom which mathematicians have in choosing the axioms of a structure. There was nevertheless an essential restriction on this freedom, and that is, as was emphasized by Cantor from 1883 onwards, the concepts introduced must not be inconsistent.

How could one be sure that this was in fact the case? Once the idea of a *model* was discovered, capable of proving the consistent nature of non-Euclidean geometries (§1, C), this idea was applied without difficulty to many other theories. The arithmetization of the theory of real numbers (§2, A), provided a model of these in the theory of natural numbers, and in consequence models of geometries in this same theory were made available. As for the various structures invented in the course of the nineteenth century, they had models in classical mathematics, since they had been introduced precisely in order to give a general formulation to examples drawn from classical theories (see Chapter V, 2).

Everything was thus finally reduced to the consistency of arithmetic. Up until the end of the nineteenth century, this was a question that was not even asked, so much did the natural numbers seem one of the most basic concepts of the human mind, and even, for some, to possess a "reality" comparable to that of the external world.

The "paradoxes" of set theory (§4, C) necessarily brought an end to this comfortable certainty; the more so because, in order to construct arithmetical "models" of other theories, it was necessary to apply to the set of natural numbers certain axioms of general set theory (Appendix 2), which threatened precisely to bring in contradictions, and to make these models useless.

It was Hilbert who, from 1904 on, proposed and developed a new method of attacking questions of consistency, and, in a more general way, of examining *a priori* what mathematical proofs can accomplish. This is what is known as *metamathematics* , or the *theory of proof* .

This method depends on the possibility of expressing the whole of mathematics in a formal language (making use of logical signs). Formalized mathematics can be compared to a game such as chess: the logical and mathematical signs are the *pieces* of the game, the axioms and theorems of a given theory are the *positions* of the pieces allowed by the rules of the game; the transition from one theorem to another following the rules of logic is a "move" allowed by the rules.

The task then is to analyse proofs by abstracting the meanings which can be attributed to them in mathematics, and to try to predict where they will end. Here again, this can be compared (not forgetting the relative proportions) with the analysis of a situation in a game of chess, for which it is possible to show either that it must necessarily end in "checkmate", or alternatively that it can never lead to this.

C. The Triumphs of Mathematical Logic

It was within the framework thus sketched by Hilbert that contemporary logicians were to obtain a series of altogether unexpected results, casting new light on the idea of set, but hardly answering to the excessively sanguine hopes of Hilbert.

The latter, with the help of several mathematicians of his school, had begun by applying his method to axiomatic systems simpler than that of Peano (§2, C), and thus incapable of encompassing all the properties of the natural numbers. For these systems of axioms, it could be shown that they did not lead to any contradiction.

In 1930, however, the young mathematician K. Gödel published a short note in which he announced that one of two things is true: either arithmetic is inconsistent or it is not, but then it is *impossible* to prove which. Thus were Hilbert's hopes reduced to nothing.

This extraordinary result can only be understood by studying in detail Gödel's technique, which is extremely complex. We can give only a very crude idea of his principle[23].

The proof proceeds along two lines of attack. Gödel began by describing a "dictionary", in which, by means of precise procedures, explicit in every detail, he associated first a natural number with each sign in a formal language, then a natural number with each combination of these signs allowed by the rules, and finally a natural number with each sequence of these combinations which constitutes a proof in accordance with the same rules.

This enabled him to describe the formation of a proposition A in arithmetic, and then to prove that, if the arithmetic is not inconsistent, then *there is no* proof *either* of A *or* of (not A) within the arithmetic. We say that A is *undecidable* within the arithmetic. Hilbert had thought that it could be proved that *any* proposition in arithmetic is "decidable" within it: Gödel's theorem is what is known as the theorem of *incompleteness* of arithmetic.

In his second line of attack, Gödel described the formation of another proposition C in arithmetic, and showed, still with the help of his "dictionary", that, if there existed a proof of C within the arithmetic, then the arithmetic would not be inconsistent. He proved, further, that the statement "C implies A" is a theorem of the arithmetic: if there existed a proof of C in the arithmetic, then there would thus also exist a proof of A, which is impossible by virtue of the incompleteness theorem, *if* the arithmetic is not inconsistent.

The undecidable proposition A described by Gödel appears to be very artificial, without any connection with any other part of the current theory of numbers. Its principal utility was to establish the impossibility of a proof of consistency in arithmetic. Among the numerous classical questions which are not resolved within number theory (see Chapter IV), it has not yet been established, as far as I know, that any of them is undecidable.

On the other hand, Gödel himself in 1940, followed by P. J. Cohen in 1963, accomplished what can be truly called a revolution in set theory. Cantor had made a conjecture which he did not know how to prove, namely, that in the set **R** of real numbers, every infinite subset is either countable or equipotent to **R**: this is what is called the *Continuum Hypothesis CH*. Many attempts were made to prove CH or to prove (not CH); but what Gödel and Cohen discovered was that, if it be accepted that ZF cannot entail a contradiction, then CH is *undecidable* within the theory of sets S, based on the axioms ZF.

---

[23] Thanks to its strangeness, this result has proved tempting to many publishers of magazines popularizing science, who have devoted many articles to it. If we add to it the four-colour theorem, Fermat's equation $x^n + y^n = z^n$ and the rules of logic used by computers, we have made a pretty well complete inventory of the mathematical questions appearing in these periodicals. One supposes that their readers believe there are no others!

The proofs are extremely technical. They are still based on the idea of *model* , used in a new way. Gödel defines a property $R(x)$ of an object x of S which expresses the fact that it can be "constructed" in a certain way, described in detail; and the *constructible* objects of S are said to be those for which $R(x)$ is a theorem of S. The theory S' has as its objects the constructible objects of S, and as its basic relation $x \in X$, but with the condition that x and X must *both* be constructible. From this relation it can be verified that the axioms of ZF are still theorems of S when they are *restricted* in the same way to the objects of S'. For example, if X is a set of S', then the subsets of X which are sets of S' are the members of a set, *of* S'. If these "restrictions" of the axioms of ZF are taken as axioms of S', then S' so defined is a *model* of S, and every relation of S is "translated" into the model S' by restricting the objects which belong to it to constructible objects.

This being the case, let CH' be the "translation" into S' of the Continuum Hypothesis CH; the relation $R(x)$ chosen by Gödel is such that CH' is a *theorem of* S', and hence CH' is also a theorem of S. If then (not CH) were a theorem of S, then by "translation", (not CH') would be a theorem of S', and hence also a theorem of S, and since CH' is also a theorem of S, S would be inconsistent. We must take care to note that this shows that there is no proof of (not CH) in S, but *does not prove* that there is a proof of CH in S.

In 1963, P. J. Cohen invented a new and remarkably flexible method of forming models of S, called "forcing". We begin with Gödel's model S', and *add* new objects of S to the objects of S'. These objects are the basic objects of a theory S'', and the basic relation $x \in X$ of this theory is defined by "restriction" as above. But this time, by a suitable definition of the objects of S'' (constituting the crucial point of Cohen's method), it is possible to bring it about that, if CH'' is the translation of CH into S'', it is (not CH'') which is a *theorem of* S'', and hence a theorem of S. We conclude from this, as above, that there is no proof of CH in S, which has the effect of establishing that CH is *undecidable in* S, provided that S is not inconsistent.

We now note, in a general way, that, if P is an undecidable proposition in S, then we can consider the axioms $ZF + P$ or $ZF + $ (not P) obtained by adding P or (not P) to Zermelo's list. If ZF does not entail a contradiction, *the same is true* of $ZF + P$ *and* of $ZF + $ (not P). Thus we have defined *two* "set theories", mutually incompatible of course, but just as valid as S. But there is more to come: the Continuum Hypothesis is still expressed by saying that there is *no* cardinality $\mathfrak{m}$ such that $\aleph_0 < \mathfrak{m} < \mathfrak{c}$. For each integer $n \in \mathbf{N}$, a hypothesis CH_n can be formulated according to which there are *exactly* n distinct cardinalities satisfying the preceding inequality (in this notation, CH is the same as CH_0). Then Cohen's method, applied each time by adding to S' basic objects suitably chosen to obtain a model S_n [24], shows that, if ZF does not entail a contradiction, then CH_n is undecidable in S,

---

[24] Two sets of S' can be non-equipotent in S' but become equipotent in S_n because the new sets added to those of S' can be such that they allow a bijection to be defined which did not exist in S'.

and it follows that $ZF + CH_n$ does not entail a contradiction either. We thus have an infinite number of "set theories", any two of which are mutually incompatible! Neither is that all that the method can do: in the last twenty years, logicians have been able to show that a number of propositions of S are undecidable, in particular the Axiom of Choice AC, and even the Axiom of Countable Choice.

D. Mathematicians' Reactions

As the axioms ZF imply Peano's axioms for arithmetic (§3, A), Gödel's incompleteness theorem shows that it is not possible, by the procedures of metamathematics, to establish that they do not entail contradictions. The mathematicians known — wrongly — as "formalists" (that means in our day almost the whole body of research mathematicians) are resigned to this. In the same way as physicists and biologists believe in the permanence of the laws of nature solely because it is compatible with what they have observed up to now, so these mathematicians are persuaded that no contradiction will come to light in set theory, since none has appeared in the last eighty years. Moreover, the same mathematicians are convinced that it is in vain to think we can have any "intuitive" concept of an infinite set: in fact, our undoubted intuition about *small* finite sets makes numbers of statements relating to infinite sets seem "natural", and yet these statements remain undecidable from the ZF axioms, which are just as "natural" from the point of view of this same intuition.

It is in protest against this resignation that mathematicians have arisen who cannot be content with a point of view in which one does not allow oneself to speak of "truth", while holding strictly to the concept of "capable of proof" in terms of a system of axioms[25]. These dissenters have never been very numerous, but they have included such eminent men as H. Poincaré, E. Borel, Lebesgue and H. Weyl. The most systematic among them has been L. E. J. Brouwer, who became noted for his first-class work in algebraic topology, but devoted the major part of his scientific career to reconstructing a mathematics in conformity with his ideas, and founded a school of mathematicians known as "intuitionists". Others, whose point of view is close to that of the intuitionists, are termed "constructivists".

It is rather difficult to describe the ideas of these mathematicians, who in any case differ one from another. They admit that mathematical objects possess a "reality" different from empirical reality (perhaps like that which Plato attributed to his "ideas"?). They are all agreed in according an exceptional status to the sequence of natural numbers, about which they think they have a basic intuition, and which they are *certain* cannot contain hidden contradictions. They do not accept the "existence" of a mathematical object unless

---

[25] The so-called "formalist" mathematicians do not refrain from speaking of "true propositions", but for them it is always a case of a proposition which has a published proof based on an explicit system of axioms. A "false proposition" is a proposition whose *negation* is "true" in the preceding sense.

they can "construct" it (§4, A), which means that sets of cardinality higher than the countable are for them only "fictions", outside the scope of mathematics. This leads them to restrict the rules of classical logic, and to refuse to make use of the so-called "law of the excluded middle", which states that for any proposition P, the proposition $(P$ or $(\text{not}\,P))$ is true, even if neither P nor $(\text{not}\,P)$ can be proved. This attitude enables them to avoid "paradoxes", but many theorems accepted by "formalists" are not accepted by "intuitionists" or "constructivists".

The controversies between the two camps were very lively in about the period 1925 – 1930, when each one was striving in vain to convince the other. Nowadays, some "constructivists" have a desire to proselytize which is scarcely reflected in the mathematical community. Young mathematicians known as "formalists" no longer even know that there was once a "crisis of foundations". This is in any case a completely misleading name, when we compare what happened in mathematics with the real crises in physics, when relativity and quantum mechanics forced the physicists radically to modify their conception of natural phenomena. In mathematics, we can at most speak of a certain uneasiness caused by the "paradoxes"; but apart from Brouwer and his disciples, there has hardly been a mathematician who was led to make the slightest change in his way of presenting proofs. Furthermore, the period of the supposed "crisis", 1895 – 1930, was one of the most fertile in history in terms of decisive advances in nearly all the items enumerated in Chapter V, 5, A. It would be even more nonsensical to speak of "crisis" at the present day, given the never-equalled abundance of solutions to ancient problems and of discoveries of new methods which we are witnessing.

E. Relations between Mathematics and Logic

Since 1930, specialists in mathematical logic have not limited their endeavours to proving the undecidability of propositions in set theory. They have also studied many logical systems different from classical logic (the one which is incorporated into the ZF system): not only the rules of logic accepted by the intuitionists, but also those of other systems, some of which stem from the philosophers of the ancient world. These are modal logics, logics of more than two values[26], etc. They have likewise analysed thoroughly the procedures of "construction" (§4, A) used by mathematicians, creating what is known as the theory of "recursive functions". The number of logicians working on these questions is continually increasing, and mathematical logic at present constitutes an imposing edifice.

Some mathematicians, using classical logic in their proofs, have drawn on the discoveries of logicians to provide answers (positive or negative) to open questions bearing especially on general topology, the theory of ordered sets, measure theory, and certain parts of algebra. From the beginning of this

---

[26] Apart from the values "true" and "false" associated with the propositions of classical logic, these logics allow a third, the "possible", and even many others.

century, they have used for this purpose the Axiom of Choice or the Continuum Hypothesis; but nowadays they can approach these problems by way of the theory whose axioms are $ZF + P$, where P is one of the many propositions undecidable by the axioms of ZF.

It has also been noticed that some new logical systems can be translated into algebraic structures, just as classical logic is translated into "Boolean algebra" (§5, A); and these structures have been studied in their own right, apart from their significance for logic.

All the same, if one looks through *Mathematical Reviews* , one notices that these works represent only a very small part of the research done by "formalists". Most of it is done in areas (see Chapter V, 5, A) where the cardinalities of sets rarely exceed that of **R**, and where it is never necessary to use any axiom outside ZF apart from the Axiom of *Countable* Choice (§4, B)[27]. Mathematicians working in these fields hardly know what the logicians are doing, and if they sometimes hear about it, they pay no more attention than they would to sciences quite remote from mathematics, such as biology or geology. If logicians sometimes show signs of surprise at this separation, it is because they have failed to take into account the evolution of mathematics over the last fifty years. Nothing that they do can help any of the mathematicians in question to solve their problems, the only ones in which they are really interested; for if they have by chance heard of the "paradoxes", as far as they are concerned they are nothing but pseudo-problems.

6. The Concept of "Rigorous Proof"

Mathematicians of every age have been seen to criticize the proofs of their predecessors or of their contemporaries as "not being rigorous"; and often those which they proposed as replacements for the defective proofs were in their turn considered inadequate by the succeeding generation. This apparent repeated putting into question of mathematicians' arguments at length led some people to think these arguments contained a hidden flaw which meant that a proof could never be found unassailable by any criticism.

I think that we have here a superficial view, due to a misunderstanding of the history of mathematics. Let us look again at the way in which a proof unfolds, conceived as a sequence of inferences: to prove a statement Q, we proceed from the fact that a prior statement P has already been proved (or is an axiom), and prove the statement "P implies Q". Since a proof is "rigorous" only if each of the inferences of which it is made up is correct, it is necessary to examine what can make an inference incorrect.

---

[27] Some of these mathematicians go so far as to declare in their writings their "repugnance" to the Axiom of Choice. Their stance is the opposite to that of the "young Turks" of 1930, who, in order to shock traditionalists, used the Axiom of Choice even when they could have done without it.

I) Leaving aside banal mistakes in calculation, a mathematician can inadvertently confuse P (respectively, Q) with an analogous proposition P' (respectively, Q'), for which P' and "P' implies Q" (respectively, P and "P implies Q'") have been proved. This is often what happens when the statement of P is long and complicated, or when the statement of Q involves a large number of cases which have to be examined separately. Sometimes it takes dozens of years to discover the mistake.

II) One of the propositions P or "P implies Q" is not an axiom and has not been proved, but appears very plausible. This is the category of insufficiency: we have pointed out some examples of it in Euclid (see Chapter III, 4). Some analogous examples are the source of errors among analysts of the early nineteenth century, such as Cauchy, Abel, and even Dirichlet and Riemann, in questions of continuity and convergence. The objects which they study are correctly defined, but they fail to check, step by step, that they are not applying anything outside these definitions. Similar cases have often occurred in the theory of algebraic surfaces right up until 1940, in which "exceptional cases", to which the generic arguments do not apply, were systematically neglected.

Errors of this type are less common in a theory already long established. In analysis, they have not occurred since the time of Weierstrass. Even today, however, it happens that a proof involves a large number of stages, some of them requiring huge efforts to be set up, while by contrast, the statement "P implies Q" appears simple, often more or less similar to standard modes of argument. The tendency is therefore to look at it rapidly, often without writing anything down: the mathematician, surfeited with details, dispenses with a written proof, replacing it with the ever-recurring phrase "It is easy to see that ... ". I do not believe there are many mathematicians who have not experienced this mischance at least once in their careers.

III) In the foregoing cases, it is only rarely that it takes much time for a proof to be rectified. The case is quite different when it is a question of purported proofs vitiated from the beginning because they are dealing with objects that are *not precisely defined* . That is what happened in analysis in the seventeenth and eighteenth centuries, when arguments took place about the "infinitely small" or the "sum" of a series, without anyone ever being able to say what this meant. Of course, in most cases the best mathematicians of the time had a just idea of what could be done with these vague notions, and their remarkable discoveries encouraged them to forge ahead; but they could not express their proofs in a properly mathematical language; and although it was not difficult, in the nineteenth century, to give entirely correct proofs of their results, this did not become possible until the idea of limit as basic concept had been clearly worked out, and its properties comprehensively codified (see Chapter III, 9).

The same situation recurred throughout the nineteenth century with respect to the elaboration of topological concepts (see Chapter V, 3, E). The visionary genius of Riemann had created a way of studying mathematically the topology of a surface, which made it possible, for example, to rationalize the "intuitive" difference between a sphere and a torus, by means of the study of curves on their surfaces. Neither he nor his contemporaries, however, possessed the technical tools indispensable for this study, and they had to fall back on "intuition" when they spoke of "curves", of "surfaces", or of "deformations": a procedure whose dangers were soon to appear in connection with Peano's "curve" (§2, B).

There is worse to come, however, even in what was considered as "elementary geometry", in relation to "polyhedrons". In his *Elements* , Euclid defines only prisms, pyramids and regular polyhedrons. In 1750, Euler spoke of "polyhedrons" without defining them, and stated that if s, a, and f are the numbers of the vertices, edges, and faces of a polyhedron P, respectively, then the number

$$\chi(P) = s - a + f$$

is always equal to 2. His proof (inadequate in any case) leads one to think that he is concerned only with *convex* polyhedrons, that is to say, with those situated on only one side of each of their faces. In the course of the nineteenth century, not only was Euler's theorem about convex polyhedrons correctly proved, but also a dozen mathematicians applied themselves to calculating the number $\chi(P)$ for *all* polyhedrons. The drawback was that, as far as I know, *none* of them was capable of giving a *general* definition of what should be understood by "polyhedron". This appeared only in Poincaré's work in 1895[28]. Previously, each of the mathematicians in question thought he had given such a definition, in accordance with his personal "intuition" of what a polyhedron must be. Shortly afterwards, another mathematician exhibited "figures" which, according to him, were also "polyhedrons", but which did not satisfy the conditions fixed by his predecessor. Naturally, the values of $\chi(P)$ depended on the "definition" chosen, so much so that it did not seem that a general expression of $\chi(P)$ would ever be arrived at which would be sufficiently "rigorous"!

The conclusion is obvious: there can be no "rigorous" proof except in the context of an axiomatic theory, in which objects and "basic" relations have been specified, and the axioms by which they are connected have been exhaustively listed; and then, if we do not count the inadvertences or negligences mentioned in I) and II), this necessary condition is also sufficient. "Lack of rigour" means nothing but "lack of precision".

---

[28] The definition is simple, but it had to be thought of! In $\mathbf{R}^3$ an *open segment* is a segment of line with its endpoints removed, an *open polygon* is a bounded, convex polygon lying in a plane of $\mathbf{R}^3$, and without its boundary. A *polyhedron* is thus a union of a *finite* set $\mathfrak{F}$ of subsets of $\mathbf{R}^3$, whose members are points, open segments and open polygons, *such that no two distinct elements of* $\mathfrak{F}$ *have a point in common* , and subject to the following condition: the endpoints of an open segment belonging to $\mathfrak{F}$ themselves belong to $\mathfrak{F}$, and the vertices and the sides (considered as open segments) of an open polygon belonging to $\mathfrak{F}$ themselves belong to $\mathfrak{F}$. Alexander proved in 1915 that for a polyhedron P, $\chi(P)$ is *invariant with respect to homeomorphism* .

History corroborates this statement in every case. There has *never* been any controversy as to what constitutes a "rigorous" proof in arithmetic; there has been none in analysis after Weierstrass; nor in algebraic topology since 1930, nor in algebraic geometry since 1950. Of course, it is not impossible that in future mathematicians will want to develop a theory without putting it into axiomatic form; but until they themselves or someone else manage to do so, the theory runs the risk of being considered "not rigorous" by the mathematical community.

VI Appendix

1. Geometry on a Surface

To abbreviate, we say that a real-valued function of one or several real variables is *regular* if it has all (partial) derivatives up to the third order.

A. Skew curves

In differential geometry, a regular skew curve Γ in the space $\mathbf{R}^3$ is the locus of a point $M(t)$ with coordinates

$$(1) \qquad x = f(t), \quad y = g(t), \quad z = h(t)$$

when t varies in an interval I of $\mathbf{R}$, and f, g, and h are regular functions on this interval. We say that t is the *parameter* of the point $M(t)$ on the curve Γ, and that the formulae (1) constitute a *parametric representation* of Γ. For example, a circular helix is defined by a parametric representation

$$(2) \qquad x = a \cos t, \quad y = a \sin t, \quad z = ht,$$

where t varies in $\mathbf{R}$ and a and h are non-zero constants.

Euler showed that the length $s(t)$ of an arc of Γ with origin a point A of the curve and endpoint $M(t)$ is a function of t having a derivative given by

$$(3) \qquad s'(t) = \left((f'(t))^2 + (g'(t))^2 + (h'(t))^2 \right)^{1/2} .$$

For example, for the helix given by (2), we have $s'(t) = (a^2 + h^2)^{1/2}$, whence

$$s(t) = (a^2 + h^2)^{1/2}\, t + \text{constant} .$$

B. Curves on a Surface

For a "regular" surface Σ embedded in $\mathbf{R}^3$, Gauss had the idea that in the same way Σ is the locus of a point $M(u, v)$ with coordinates

$$(4) \qquad x = f(u, v), \quad y = g(u, v), \quad z = h(u, v),$$

where u and v each vary in an interval of $\mathbf{R}$, and f, g, and h are regular functions. A regular curve Γ traced on Σ is then obtained by taking for u

and v two regular functions of t on an interval of $\mathbf{R}$, and by substituting these functions for u and v in (4). The formula (3) then gives

$$(s'(t))^2 = E(u'(t))^2 + 2Fu'(t)v'(t) + G(v'(t))^2$$

for the length $s(t)$ of such a curve, where E, F, and G are functions of u and v defined by

(6)
$$E(u,v) = \left(\frac{\partial f}{\partial u}\right)^2 + \left(\frac{\partial g}{\partial u}\right)^2 + \left(\frac{\partial h}{\partial u}\right)^2 ,$$

$$F(u,v) = \frac{\partial f}{\partial u}\frac{\partial f}{\partial v} + \frac{\partial g}{\partial u}\frac{\partial g}{\partial v} + \frac{\partial h}{\partial u}\frac{\partial h}{\partial v} ,$$

$$G(u,v) = \left(\frac{\partial f}{\partial v}\right)^2 + \left(\frac{\partial g}{\partial v}\right)^2 + \left(\frac{\partial h}{\partial v}\right)^2 ,$$

and one must substitute for u and v their expressions in terms of t in this formula. It is convenient to use Leibniz's notation

$$\frac{d\varphi}{dt}$$

for the derivative of a function $\varphi(t)$, and then formula (5) can be written in the form

$$ds^2 = E\,du^2 + 2\,F\,du\,dv + G\,dv^2 ,$$

which has the advantage of not specifying the chosen parameter t.

If two curves Γ_1 and Γ_2 traced on Σ pass through the same point, the *angle* α between their tangents at this point, suitably oriented, is given by

(8) $\cos \alpha =$

$$\frac{E\,du_1\,du_2 + F(du_1\,dv_2 + dv_1\,du_2) + G\,dv_1\,dv_2}{\left(E\,du_1^2 + 2\,F\,du_1\,dv_1 + G\,dv_1^2\right)^{1/2}\left(E\,du_2^2 + 2\,F\,du_2\,dv_2 + G\,dv_2^2\right)^{1/2}} .$$

The geodesics of Σ are defined as the curves $v = \varphi(u)$, where φ satisfies a certain second order differential equation depending only on the functions E, F, and G, and their derivatives.

Finally, if a set Δ on Σ corresponds to points (u,v) of a subset D of the plane $\mathbf{R}^2$, then we have

(9)
$$\text{area}\,\Delta = \iint_D (EG - F^2)^{1/2}\,du\,dv .$$

It is necessary to remark that it is not always possible to find a parametric representation (4) for *the whole* of a surface Σ such that each point of Σ corresponds to *only one* pair (u,v) of parameters. For example, for the sphere with centre O and radius R, we have the following parametric representation as a function of longitude φ and latitude θ:

$$x = R\cos\theta\cos\varphi, \quad y = R\cos\theta\sin\varphi, \quad z = R\sin\theta,$$

where $0 \leq \varphi \leq 2\pi$ and $-\pi/2 \leq \theta \leq \pi/2$. But the values $\varphi = 0$ and $\varphi = 2\pi$ correspond to the same point, and the value $\theta = \pi/2$ all give the north pole of the sphere for every value of φ. It is thus necessary to apply the preceding formulae only to the *portion* of the sphere defined by

$$0 < \varphi < 2\pi, \quad -\frac{\pi}{2} < \theta < \frac{\pi}{2}.$$

Riemann's point of view consisted in forgetting any sort of "embedding" of a surface in space, and in *defining* a "portion of a surface" as a product of two open intervals $a < u < b$ and $c < v < d$ of **R**, but with a definition of "length" of curves which in general is not that of the Euclidean plane. The length is given by formula (7), but now it is supposed *only* that E, F, and G are regular functions of (u, v) such that $E > 0$, $G > 0$, and $EG - F^2 > 0$ *without deducing them* from formulae such as (6).

Example: The Flat Torus. A torus is the surface of revolution generated by a circle of radius r turning around a line in its plane at a distance $a > r$ from its centre. Taking the axis of rotation as the axis Oz, we have a parametric representation

$$x = (a + r \cos u) \cos v, \quad y = (a + r \cos u) \sin v, \quad z = r \sin u,$$

where $0 \leq u \leq 2\pi$ and $0 \leq v \leq 2\pi$.

If the lengths of curves on this surface are calculated by Euler's formula (3), then we obtain

$$ds^2 = (a + r \cos u)^2 \, dv^2 + r^2 \, du^2.$$

On the other hand, for the "flat torus", we take simply

$$ds^2 = du^2 + dv^2,$$

with the result that the portion of surface defined by $0 < u < 2\pi$ and $0 < v < 2\pi$ is not distinguished from this subset of the plane, in accordance with Riemann's point of view.

C. Poincaré's Half-Plane

We now consider the surface which is the half-plane H formed of the points (x, y), where $x \in \mathbf{R}$ is arbitrary and $y > 0$. We take

(10) $$ds^2 = \frac{1}{y^2}(dx^2 + dy^2).$$

A semicircle with centre on Ox is given by the parametric representation

$$x = a + r \cos \varphi, \quad y = r \sin \varphi,$$

where $0 < \varphi < \pi$, and it then follows from (10) that the length of this curve is given by

$$(11) \qquad ds = \frac{d\varphi}{\sin \varphi} \,,$$

whence

$$s = \log \tan \frac{\varphi}{2} + \text{constant} \,.$$

For the "non-Euclidean" angle between two semicircles (see Figure 55), formula (8) gives

$$\cos \alpha = \cos (\varphi_2 - \varphi_1)$$

because of (11). The "non-Euclidean" angle α is thus *equal* to the "Euclidean" angle $\varphi_2 - \varphi_1$.

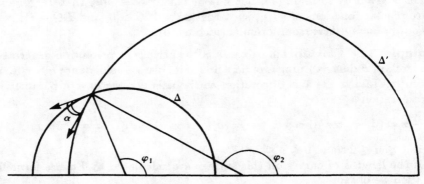

Fig. 55.

To study the action of the group $GL^+(2, \mathbf{R})$, it is convenient to write formula (10) in the form

$$(12) \qquad ds^2 = \frac{1}{y^2} \, dz \cdot d\bar{z} \,.$$

If now

$$Z = X + iY = \frac{\alpha z + \beta}{\gamma z + \delta} \,,$$

then we have

$$\bar{Z} = \frac{\alpha \bar{z} + \beta}{\gamma \bar{z} + \delta} \,,$$

whence

$$dZ = \frac{\alpha \delta - \beta \gamma}{|\gamma z + \delta|^2} \, dz \,, \quad d\bar{Z} = \frac{\alpha \delta - \beta \gamma}{|\gamma z + \delta|^2} \, d\bar{z} \,,$$

and

$$Y = \frac{\alpha \delta - \beta \gamma}{|\gamma z + \delta|^2} \, y \,,$$

and so

$$\frac{1}{Y^2} \, dZ \cdot d\bar{Z} = \frac{1}{y^2} \, dz \cdot d\bar{z} \,.$$

For the "transitivity" of this group on the "segments" of the "non-Euclidean" line of the same (non-Euclidean) length, we notice that the group $GL^+(2, \mathbf{R})$ contains the subgroup of homotheties and translations

$$z \mapsto \alpha z + \beta \, ,$$

where $\alpha > 0$ and β is any real number.

Using this group, it is sufficient to show that each arc $\overset{\frown}{A_1 A_2}$ of the semicircle $x^2 + y^2 - 2x = 0$ can be transformed by an element of $GL^+(2, \mathbf{R})$ into a segment of the line $B_1 B_2$ parallel to Oy of the same non-Euclidean length. But the transformation

$$Z = 4 - \frac{4}{z}$$

transforms the point (x, y) of the semicircle to the point (X, Y) with $X = 2$ and $Y = 2y/x$, and we have (Figure 56):

$$\log Y_2 - \log Y_1 = \log \tan \frac{\varphi_2}{2} - \log \tan \frac{\varphi_1}{2} \, .$$

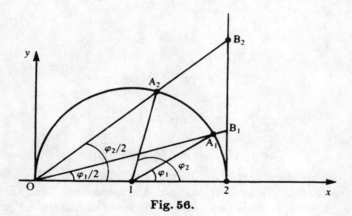

Fig. 56.

This transitivity can be used to calculate the area of a "non-Euclidean" triangle ABC. Since $GL^+(2, \mathbf{R})$ conserves angles and areas, it is sufficient to carry out the calculation in the case where BC is part of a line parallel to Oy (Figure 57). We have then by (9) that

$$\text{area}\,(ABC) = \int_p^q dx \int_{r \sin \varphi}^{R \sin \psi} \frac{dy}{y^2}$$

$$= \int_p^q \frac{dx}{r \sin \varphi} - \int_p^q \frac{dx}{R \sin \psi} \, .$$

Now, on the arc of the circle $\overset{\frown}{CA}$, we have $x = r \cos \varphi$ and on the arc of the circle $\overset{\frown}{BA}$ we have $x = R \cos \psi$, whence

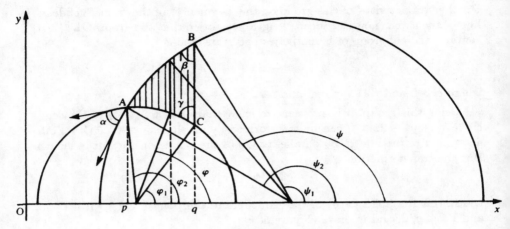

Fig. 57.

$$\int_p^q \frac{dx}{r \sin \varphi} = \varphi_2 - \varphi_1 , \quad \int_p^q \frac{dx}{R \sin \psi} = \psi_2 - \psi_1 .$$

Thus

$$\text{area}(ABC) = \varphi_2 - \varphi_1 - (\psi_2 - \psi_1)$$
$$= (\varphi_2 - \psi_2) - \varphi_1 + \psi_1 .$$

But

$$\alpha = \psi_2 - \varphi_2 , \quad \gamma = \varphi_1 , \quad \text{and} \quad \pi - \beta = \psi_1 ,$$

and so, finally,

$$\text{area}(ABC) = \pi - (\alpha + \beta + \gamma) .$$

Finally, let us show that there exists only a *single* non-Euclidean line which is perpendicular to two "parallel" non-Euclidean lines. In the Euclidean "dictionary", these two "lines" are two semicircles of diameters AA' and BB' which have no common point (Figure 58).

If these semicircles are concentric, the unique "common perpendicular" is the half-line parallel to Oy through their common centre. Otherwise, denote by C the centre and by R the radius of a semicircle with centre on Ox, orthogonal to the two given semicircles. If x, a, a', b, b' are the abscisae of C, A, A', B, B', respectively, then we must have

$$R^2 = \overline{CA} \cdot \overline{CA'} = \overline{CB} \cdot \overline{CB'} ,$$

so that

(13) $$R^2 = (x - a)(x - a') = (x - b)(x - b') .$$

This gives x as the solution of the first-degree equation

$$(a + a' - b - b')x = bb' - aa' ,$$

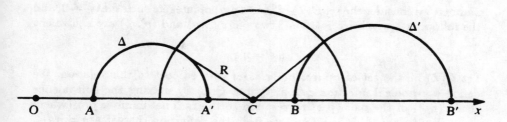

where the coefficient $a + a' - b - b'$ is not zero because the semicircles are not concentric. This value of x when inserted into (13) must give a positive value to the two expressions on the right-hand side, for otherwise there would be a positive real number y such that

$$(x - a)(x - a') = (x - b)(x - b') = -y^2$$

and this would imply that the two semicircles

$$(x - a)(x - a') + y^2 = 0, \quad (x - b)(x - b') + y^2 = 0$$

had a point in common in the half-plane H, contrary to the hypothesis.

2. Models of the Real Numbers

A. The Theory of Rational Numbers

The axioms for the real numbers (see Chapter III, Appendix 2) imply the existence within $\mathbf{R}$ of the additive subgroup $\mathbf{Z}$ of the rational integers and of the subfield $\mathbf{Q}$ of the rational numbers p/q, where $p \in \mathbf{Z}$ and $q \in \mathbf{Z}$ with $q \neq 0$. To obtain *models* of the real numbers in the theory of natural numbers, we first form *models* of $\mathbf{Z}$ and of $\mathbf{Q}$ in this theory.

For $\mathbf{Z}$, the following *equivalence relation* in $\mathbf{N}^* \times \mathbf{N}^*$ is considered: two pairs (p, q) and (p', q') are equivalent if

$$p + q' = p' + q,$$

and we denote by $Z(\mathbf{N}^*)$ the set of *equivalence classes* for this relation.

There is then a canonical *bijection* d of $\mathbf{Z}$ onto $Z(\mathbf{N}^*)$: to define it, we let the pair $(z + 1, 1)$ correspond to $z \in \mathbf{Z}$ if $z \geq 0$, and let $(1, 1 - z)$ correspond to z if $z < 0$ (we recall that we have identified $\mathbf{N}^*$ with the set of real numbers $n \cdot 1$, where $n \in \mathbf{N}^*$). Then $d(z)$ is the equivalence class of this pair. We have the model that we sought; the verifications are immediate.

For $\mathbf{Q}$, we form a model in the theory of integers (and, from this, if we wish, a model in the theory of natural numbers), by proceeding in an analogous manner: we consider the set $C(\mathbf{Z})$ of pairs (p,q) of integers such that $q \neq 0$, and the following *equivalence relation* : two pairs (p,q) and (p',q') are equivalent if

$$pq' = p'q.$$

Let $Q(\mathbf{Z})$ be the set of *equivalence classes* with respect to this relation. We define a canonical *bijection* c from $\mathbf{Q}$ into $Q(\mathbf{Z})$ by making the equivalence class $c(p/q)$ of the pair (p,q) correspond to the rational number p/q (where $p,q \in \mathbf{Z}$ with $q \neq 0$). It is immediate from the definition that, if $p/q = p'/q'$, then $c(p/q) = c(p'/q')$, and so we have thus defined the sought-for model.

B. The Model of Dedekind (Simplified Account)

The model in question is a model of $\mathbf{R}$ in the theory of the rational numbers (and hence, as a consequence of A), a model in the theory of natural numbers). In the set $\mathfrak{P}(\mathbf{Q})$ of all the subsets of the set of rational numbers, we define a subset $\mathfrak{R}(\mathbf{Q})$ formed of the subsets $X \subset \mathbf{Q}$ such that $X \neq \emptyset$, $X \neq \mathbf{Q}$, and such that X satisfies the following property:

X does not have a maximum element, and, if $x \in X$ and $y < x$, then $y \in X$.

To abbreviate, we say that the elements of $\mathfrak{R}(\mathbf{Q})$ are the *segments* of $\mathbf{Q}$.

The set $\mathfrak{R}(\mathbf{Q})$ is given the structure of *order* defined by the inclusion $X \subset Y$. On $\mathfrak{R}(\mathbf{Q})$ this is a *totally ordered* structure. In fact, let us suppose that X and Y are two distinct segments. Then either there exists $x \in X$ such that $x \notin Y$ or there exists $y \in Y$ such that $y \notin X$. In the first case, the rational numbers $u \geq x$ cannot belong to Y, for otherwise by definition we would have $x \in Y$, and so *every* $y \in Y$ satisfies the condition that $y < x$, and this implies by definition that $y \in X$, whence $Y \subset X$. By the same argument in the other case we have $X \subset Y$.

For every *real* number x, we define S_x to be the segment consisting of the *rational* numbers $y < x$. The mapping $x \mapsto S_x$ is then a bijection of $\mathbf{R}$ onto $\mathfrak{R}(\mathbf{Q})$, and it is an *isomorphism* of ordered sets. In fact, if $x \neq y$ in $\mathbf{R}$, and if, for example, $x < y$, then there exists a rational number r such that $x < r < y$ as a consequence of Archimedes' Axiom (Chapter III, Appendix 2), and hence we have $S_x \subset S_y$ and $S_x \neq S_y$. On the other hand, if $S \in \mathfrak{R}(\mathbf{Q})$ is an *arbitrary* segment, then there is a rational number r such that $S \subset S_r$, for otherwise we would have $S = \mathbf{Q}$. As a consequence of the Axiom of Nested Intervals, there exists *in* $\mathbf{R}$ a least upper bound x of S (its *supremum*), and hence we have $S = S_x$. We have succeeded in giving a model $\mathfrak{R}(\mathbf{Q})$ of $\mathbf{R}$ in the theory of rational numbers.

C. The Model of Méray-Cantor (Simplified Account)

We consider the set, denoted by $\mathbf{Q^N}$, of *sequences* $n \mapsto a_n$ (maps from $\mathbf{N}$ into $\mathbf{Q}$) of sequences of rational numbers. In this set, a subset $\mathfrak{C}(\mathbf{Q})$ formed of the sequences with the following property is defined:

for each natural number p, there exists a natural number $c(p)$ such that $|a_m - a_n| \leq 1/p$ for each pair (m, n) of integers such that $m \geq c(p)$ and $n \geq c(p)$.

These elements of $\mathfrak{C}(\mathbf{Q})$ are called *Cauchy sequences* of rational numbers. An *equivalence relation* is defined in $\mathfrak{C}(\mathbf{Q})$: two Cauchy sequences (a_n) and (b_n) are equivalent if the sequence $(a_n - b_n)$ tends to 0 in $\mathbf{Q}$. Let $\mathfrak{C}'(\mathbf{Q})$ be the set of *equivalence classes* of $\mathfrak{C}(\mathbf{Q})$ for this relation: there is then a canonical *bijection* s from $\mathbf{R}$ onto $\mathfrak{C}'(\mathbf{Q})$. To define this bijection, we say that, to each $x \in \mathbf{R}$, there corresponds a sequence (a_n) of rational numbers which has x as a limit. Such a sequence exists by the Axiom of Nested Intervals and Archimedes' Axiom. All these sequences belong to $\mathfrak{C}(\mathbf{Q})$ and are *equivalent*, and we take for $s(x)$ their equivalence class. It is immediate that, if $x \neq y$ in $\mathbf{R}$, then a sequence (a_n) tending to x cannot be equivalent to a sequence (b_n) tending to y because the sequence $(a_n - b_n)$ tends to $x - y \neq 0$. On the other hand, each Cauchy sequence (a_n) has a limit $x \in \mathbf{R}$ as a consequence of the Axiom of Nested Intervals, and $s(x)$ is the equivalence class of (a_n).

3. Theorems of Cantor and of his School

A. The Set of Real Numbers is not Countable

The proof is by contradiction, supposing that there is a bijection $n \mapsto a_n$ of $\mathbf{N}$ onto $\mathbf{R}$. We define by recurrence a map $n \mapsto p(n)$ of $\mathbf{N}$ into itself in the following manner.

1) $p(0) = 0$.

2) $p(1)$ is the smallest integer n such that $a_n > a_0$.

3) Let us suppose that $p(n)$ has been defined for $n \leq 2m - 1$ and that the inequality

$$a_{p(2m-2)} < a_{p(2m-1)}$$

is satisfied. Then the set of real numbers x such that

$$a_{p(2m-2)} < x < a_{p(2m-1)}$$

is infinite; $p(2m)$ is then the smallest integer k such that $k > p(2m - 1)$ and

$$a_{p(2m-2)} < a_k < a_{p(2m-1)}.$$

4) In the same way, $p(2m+1)$ is the smallest integer k such that $k > p(2m)$ and

$$a_{p(2m)} < a_k < a_{p(2m-1)}.$$

It is clear, by definition, that $p(n+1) > p(n)$ for each $n \geq 0$. By induction, it follows that $p(n) \geq n$ for each $n \geq 0$. We have thus defined a sequence

$$[a_{p(2m)}, a_{p(2m+1)}]$$

of closed intervals such that each of these intervals is contained in the preceding one. More precisely, we have

$$(1) \qquad a_{p(2m-2)} < a_{p(2m)} < a_{p(2m+1)} < a_{p(2m-1)}$$

for each $m \geq 1$. As a consequence of the Axiom of Nested Intervals, there exists a real number y belonging to all these intervals, and as a consequence of (1), it cannot coincide with the endpoints of any of them.

By hypothesis, there exists an integer q such that $y = a_q$. Since the sequence $(p(n))$ increases indefinitely, there exists a largest integer n such that $p(n) \leq q$, and hence $q < p(n+1)$. Suppose first that n is even, say $n = 2m$. We would then have

$$a_{p(2m)} < a_q < a_{p(2m+1)} < a_{p(2m-1)},$$

and hence $p(2m+1)$ is not *the smallest* integer k such that $k > p(2m)$ and $a_{p(2m)} < a_k < a_{p(2m-1)}$. Suppose on the other hand that n is odd, say $n = 2m - 1$. We would then have

$$a_{p(2m-2)} < a_{p(2m)} < a_q < a_{p(2m-1)},$$

and hence $p(2m)$ would not be *the smallest* integer k such that $k > p(2m-1)$ and $a_{p(2m-2)} < a_k < a_{p(2m-1)}$. In both cases we have certainly been led to a contradiction.

B. The Order Relation among Cardinals

Let E and F be two *arbitrary* non-empty sets, let $f : E \to F$ be a bijection of E onto a subset $f(E) \subset F$, and let $g : F \to E$ be a bijection of F onto a subset $g(F) \subset E$. The problem is to prove that E and F are *equipotent*.

We set $h = g \circ f : E \to E$, so that h is a bijection of E onto a subset $h(E) \subset E$. Let $R = E \setminus g(F)$, and let us consider the property

$$M \supset R \cup h(M)$$

(where the right-hand side denotes the union of R and of $h(M)$) for an element $M \in \mathfrak{P}(E)$; let $\mathfrak{F} \subset \mathfrak{P}(E)$ be the set of elements $M \in \mathfrak{P}(E)$ having this property. The set $\mathfrak{F}$ is not empty because clearly $E \in \mathfrak{F}$. Let A be the *intersection* of all the sets $M \in \mathfrak{F}$. This set also belongs to $\mathfrak{F}$. In fact, we have $R \subset A$ whence $R \subset M$ for all $M \in \mathfrak{F}$; morover, $h(A) \subset h(M)$ for each $M \in \mathfrak{F}$, and so

$$R \cup h(A) \subset R \cup h(M) \subset M$$

for each $M \in \mathfrak{F}$, which implies that $R \cup h(A) \subset A$ by definition.

Secondly, we show that $A = R \cup h(A)$. Arguing by contradiction, suppose that there exists $x \in A$ not belonging to $R \cup h(A)$. Set $M = A \setminus \{x\}$. Since $x \notin h(A)$, we have

$$h(A) \subset A \setminus \{x\} = M,$$

and on the other hand $h(M) \subset h(A)$. We would then have $R \cup h(M) \subset M$, a contradiction of the definition of A.

Let $A' = f(A)$, and set $B' = F \setminus f(A)$ and $B = g(B')$. We remark that $g(A') = h(A)$. Since g is a bijection of F onto $g(F)$ and since $A' \cap B' = \emptyset$, we have $B \cap g(A') = \emptyset$, that is to say, $B \cap h(A) = \emptyset$. Since $B \cap R = \emptyset$, we have $B \cap A = \emptyset$.

Let us show that $B = E \setminus A$, the complementary set to A. In fact, take $x \in E \setminus A$. By definition, there exists $M \in \mathfrak{F}$ such that $x \notin M$, and *a fortiori* $x \notin h(M)$. On the other hand $x \notin R$, which means that $x \in g(F)$. Let $y \in F$ be the unique element such that $x = g(y)$; we show that necessarily $y \in B'$, whence $x \in B$. In the contrary case, we would have

$$y \in A' = f(A),$$

in other words $y = f(u)$ for an element $u \in A$, and hence

$$x = g(f(u)) = h(u) \in h(A).$$

But since $h(A) \subset h(M)$, we would have $x \in h(M)$. However, we have shown that $x \notin h(M)$ (Figure 59).

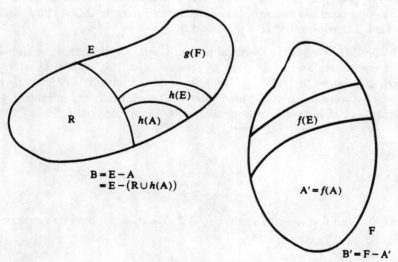

$$B = E - A$$
$$= E - (R \cup h(A))$$

Fig. 59.

We have thus decomposed E into a union $A \cup B$ of two subsets *with no common element*, and F into a union $A' \cup B'$ of two subsets *with no common element*. Further, the restriction f_1 of f to A is a bijection of A onto A', and the restriction g_1 of g to B' is a bijection of B' onto B. On taking $\Phi(x) = f_1(x)$ for $x \in A$ and $\Phi(x) = g_1^{-1}(x)$ for $x \in B$, we have clearly defined a *bijection Φ from E onto F*.

We remark that the above proof does not use the existence of an infinite set.

C. The Equipotence of R and $R^2 = R \times R$

The proof is carried out in four steps.

I) The set R is equipotent to the open interval I of numbers t such that $0 < t < 1$. For this, it is sufficient to notice that

$$t \mapsto \tan \pi \left(t - \frac{1}{2} \right)$$

is a bijection from I onto R.

II) The set I is equipotent to the subset $J = I \setminus (I \cap Q)$ of *irrational* numbers belonging to I. We use the result of B, above. In the first place, we have $J \subset I$. On the other hand, let θ be an irrational number such that $1 < \theta < 2$. The set A of numbers $t + \theta$, where $t \in I \cap Q$, is formed of *irrational* numbers and is contained in the open interval $0 < t < 3$, and so also in the set $3J$ of irrational numbers belonging to this interval. But $t \mapsto 3t$ is a bijection of J onto $3J$, and, on the other hand, I is the union of the sets J and $I \cap Q$ which have no common element, and so I is equipotent to $J \cup A \subset 3J$, and it follows that I is equipotent to a subset of J.

III) The third step is the essential part of the proof. It consists in showing that J is equipotent to the set S of *all sequences* (a_n) *of natural numbers* (a set which is denoted by $(N^*)^{N^*}$). The bijection of J onto S is defined by the theory of *continued fractions*. It associates with each irrational number $x \in J$ a sequence (a_n) of natural numbers defined by recurrence on n. Since

$$\frac{1}{x} > 1,$$

we can write

$$\frac{1}{x} = a_1 + x_1$$

with $a_1 \in N^*$ and $0 < x_1 < 1$, so that $x_1 \in J$. We then proceed by recurrence, setting

$$\frac{1}{x_1} = a_2 + x_2$$

$$\cdots\cdots\cdots\cdots$$

$$\frac{1}{x_{n-1}} = a_n + x_n$$

$$\cdots\cdots\cdots\cdots,$$

where the a_n are natural numbers and $0 < x_n < 1$ for each n. The recurrence will never come to an end, for none of the x_n can be rational: if x_n were rational, then x_{n-1}, x_{n-2}, $\ldots$, x_1 would be rational, and finally x would be rational, contrary to the hypothesis. Thus we have defined for each $x \in J$ an infinite sequence $s(x) = (a_n)_{n \geq 1}$ of natural numbers.

We have to prove that, for *every* sequence $(a_n)_{n \geq 1}$, there exists *one and only one* $x \in J$ such that $s(x) = (a_n)_{n \geq 1}$. This is what Euler did by the following argument. We first consider, for each n, the partial finite sequence $(a_1, a_2, \ldots, a_n)$, and we define a finite sequence of *rational* numbers belonging to I by the conditions

$$\frac{1}{y_1} = a_n$$

$$\frac{1}{y_2} = a_{n-1} + y_1$$

(2)
$$\cdots\cdots\cdots\cdots\cdots$$

$$\frac{1}{y_{n-1}} = a_2 + y_{n-2}$$

$$\frac{1}{y_n} = a_1 + y_{n-1} \, .$$

(We note that, if $y_n = u/v$, where u and v are two natural numbers with $u < v$, then the above equations (2), read from the bottom to the top, are none other than the sequence of successive Euclidean divisions for the calculation of the greatest common divisor of u and v in another notation (see Chapter IV, Appendix 3). Indeed,

$$\begin{aligned}
v &= a_1 u + r_1 & r_1 &< u, \\
u &= a_2 r_1 + r_2 & r_2 &< r_1, \\
r_1 &= a_3 r_2 + r_3 & r_3 &< r_2, \\
&\cdots\cdots\cdots\cdots & &\cdots\cdots\cdots \\
r_{n-3} &= a_{n-1} r_{n-2} + r_{n-1} & r_{n-1} &< r_{n-2}, \\
r_{n-2} &= a_n r_{n-1} & r_{n-1} &= \{u, v\},
\end{aligned}$$

where $\{u, v\}$ is the greatest common divisor of u and v. The natural numbers a_j are those which occur in (2), and the y_j are given by

$$y_{n-j} = \frac{r_j}{r_{j-1}}$$

for $1 \leq j \leq n - 1$.)

The number y_n is denoted by

(3)
$$\frac{1|}{|a_1} + \frac{1|}{|a_2} + \ldots + \frac{1|}{|a_n},$$

and we say that a_1, a_2, $\ldots$, a_n are the *partial quotients* of y_n, and (3) is called the *development as a continued fraction* of y_n. It is shown that, if the natural numbers p_j and q_j are defined by recurrence for $1 \leq j \leq n$ by the formulae

$$(4) \qquad \begin{aligned} &p_1 = 1, \quad q_1 = a_1, \quad p_2 = a_2, \quad q_2 = 1 + a_2 a_1, \\ &p_j = p_{j-2} + a_j p_{j-1}, \quad q_j = q_{j-2} + a_j q_{j-1} \quad \text{for} \quad j \geq 3, \end{aligned}$$

then we have $y_n = p_n / q_n$. If we return to the infinite sequence $(a_n)_{n \geq 1}$, then the p_j and q_j are defined by recurrence for *each* natural number j. Euler showed by induction that, for each $n \geq 2$, we have

$$(5) \qquad \frac{p_n}{q_n} - \frac{p_{n-1}}{q_{n-1}} = \frac{(-1)^{n-1}}{q_n q_{n-1}},$$

and it follows from (4) by induction, starting from the fact that $a_n \geq 1$ for all n, that we have $q_n \geq n$ for all n. Thus

$$\left| \frac{p_n}{q_n} - \frac{p_{n-1}}{q_{n-1}} \right| \leq \frac{1}{n(n-1)}.$$

Since the series with general term $1/n(n-1)$ is convergent, the sequence (p_n/q_n) tends to a limit x. Moreover, as a consequence of (5), for $m \geq 1$, the closed intervals

$$\left[\frac{p_{2m}}{q_{2m}}, \frac{p_{2m+1}}{q_{2m+1}} \right]$$

are nested and are contained in I, and so $x \in I$. The number x is denoted by the expression

$$\frac{1|}{|a_1} + \frac{1|}{|a_2} + \ldots + \frac{1|}{|a_n} + \ldots,$$

called the *development as a continued fraction* (unlimited) of x, the rational numbers p_n/q_n being called the *approximants* of this development. We see also that, if

$$x_1 = \frac{1|}{|a_2} + \frac{1|}{|a_3} + \ldots + \frac{1|}{|a_n} + \ldots,$$

then we have

$$\frac{1}{x} = a_1 + x_1.$$

By recurrence, it is concluded that x is irrational and that the sequence $(a_n)_{n \geq 1}$ of partial quotients of x is certainly the sequence $s(x)$.

IV) We thus reach the point where we have to show that the set $S \times S$ is *equipotent* to S. For this, Cantor simply observed that the mapping

$$((b_n), (c_n)) \mapsto (a_n),$$

where $a_{2k-1} = b_k$ and $a_{2k} = c_k$ for all $n \geq 1$, is a bijection from $S \times S$ onto S.

D. The Cardinal of the Set of Subsets

Let E be an *arbitrary* set. The mapping $x \mapsto \{x\}$, which makes the subset $\{x\}$ of E – this is the element of $\mathfrak{P}(E)$ whose only member is x [1] – correspond to each element $x \in E$, is a bijection from E onto a subset of $\mathfrak{P}(E)$. We have to see that there *does not exist* any bijection of E onto $\mathfrak{P}(E)$. We argue by contradiction from the assumption that there is such a mapping $f : E \to \mathfrak{P}(E)$. Let us consider the property $x \notin f(x)$ for elements $x \in E$. By Zermelo's Axiom of Comprehension (§4, C), there exists a subset $X \subset E$ formed of those $x \in E$ having this property. We show that $X \notin f(E)$, which contradicts the hypothesis that $f(E) = \mathfrak{P}(E)$. But, if $x \in X$, we have by definition that $x \notin f(x)$, whence $f(x) \neq X$. On the other hand, if $x \notin X$, we have by definition that $x \in f(x)$, whence again $f(x) \neq X$. Thus X cannot be an element of $f(E)$.

We remark that, in the above proof, we have used neither the Axiom of Infinity nor the Axiom of Choice.

When E is a finite set of n elements, it is proved by induction on n that $\mathfrak{P}(E)$ has 2^n elements, which explains the general notation $2^{\mathfrak{m}}$ for an arbitrary cardinal $\mathfrak{m}$.

---

[1] The distinction between an object x and the set $\{x\}$ formed of the single element x (which is due to Frege) may appear to be pedantic, but it shows itself to be useful in arguments in set theory, of which this is only a very simple example. In more complicated proofs, it is important to always know which set contains the objects that are being studied, in order to avoid confusion.

VII Index

1. Historical Index

The reader will find in this index some information about the mathematicians cited in this work. We have in principle confined ourselves to those who died before 1985. For far more complete and detailed information, the reader may consult the *Dictionary of Scientific Biography* , 16 volumes, New York (Ch. Scribner's Sons), 1970 – 1980.

ABEL, Niels Henrik (1802 – 1829)

Abel was born on the Norwegian island of Finnöy, and studied at the University of Oslo. In 1825 he received a government grant which enabled him to travel to Berlin and Paris. He died of tuberculosis at Froland in Norway. He was one of the originators of the theory of elliptic functions, independently of Gauss and Jacobi, and he founded the theory of Abelian integrals. Also owing to him are numerous theorems in analysis and the discovery of algebraic equations (now called "abelian") which are solvable by radicals.

D'ALEMBERT, Jean le Rond (1717 – 1783)

D'Alembert was born in Paris, the illegitimate son of two members of the nobility. He studied law, medicine, and mathematics at the Collège des Quatre Nations. In 1741, he was elected member of the Académie des Sciences, and, in 1772, secretary for life of the Académie Française. During the 1750s he was the scientific editor of Diderot's *Encyclopédie* . He left many works on analysis and on mechanics.

ALEXANDER, James Waddell (1888 – 1971)

Alexander was born at Sea Bright, New Jersey. He taught at the University of Princeton from 1928 to 1933 and, from 1933, at the Institute for Advanced Study at Princeton. He was the author of important works on algebraic topology.

AL-KHWĀRIZMĪ, Abū Ja'far Muhammad Ibn Mūsā (born before 800, died after 847)

Al-Khwārizmī was probably born in Iran, and was a member of a company of scientists founded by the Abassid caliphs. He was the author of writings on astronomy and of a treatise on algebra which contained nothing original, but which had very great influence during the Middle Ages.

AMPERE, André-Marie (1775 – 1836)

Ampère was born at Lyons, and taught mathematics there at the beginning of his career. In 1802 he became a teacher at the Ecole Centrale of Bourg - en - Bresse, in 1803, a junior lecturer at the Ecole Polytechnique, and in 1824, professor of experimental physics at the Collège de France. He was a member of the Institute from 1814. He is best known for his works on physics.

APOLLONIUS (second half of the third century B.C. to beginning of the second century)

Apollonius is said to have been born at Perga, a small Greek village in Asia Minor. Little is known of his life. He lived at Alexandria, and visited Pergamon and Ephesus. He was the author of a very full treatise on conics.

ARCHIMEDES (287 – 212 B.C.)

Archimedes lived at Syracuse, and very probably visited Alexandria, where he worked with Euclid's successors. In any case, he returned to Syracuse, where he was in charge of the port installations and naval and military constuction work, and he was killed during the siege of the town by the Romans. His work is the most important of classical times, and is a direct precursor of the infinitesimal calculus.

ARGAND, Jean Robert (1768 – 1822)

Argand was born in Geneva, and spent his life as an accountant in Paris. We know almost nothing about him. He was self-taught in mathematics.

BANACH, Stefan (1892 – 1945)

Banach was born in Cracow, in Poland, and studied at the Institute of Technology in Lvov (Ukraine), where he became professor in 1927. He was one of the founders of functional analysis, and he proved some of its most important theorems.

BELTRAMI, Eugenio (1835 – 1899)

Beltrami was born at Cremona, in Italy, and received his university education at Pavia and at Milan. He was a professor at the Universities of Bologna (1862 – 64, 1866 – 73), of Pisa (1864 – 66), of Rome (1873 – 76, 1891 – 99) and of Pavia (1876 – 91). He became President of the Accademia dei Lincei in 1898. His most important writings are on differential geometry.

BERNOULLI, Daniel (1700 – 1782)

Daniel Bernoulli was born at Groningen in the Netherlands, where his father, John Bernoulli, was a professor. From 1705 he lived at Basel, apart from a stay in Venice (1723 – 24) and eight years spent in St. Petersburg (1725 – 33), where he worked at the Academy of Sciences. From 1733 to 1776 he was professor, first of botany and anatomy, then of physics, at the University of Basel. His most important writings are about hydrodynamics, probability theory, and the equation of vibrating strings.

BERNOULLI, Jacob (1654 – 1705)

Jacob Bernoulli was born at Basel, and studied philosophy, theology, mathematics and astronomy at the University of his native town, the two latter against his father's will. After study tours in France, the Netherlands and England, he taught from 1683 at the University of Basel, and mathematics in particular after 1687. A disciple of Leibniz, he developed numerous applications of the infinitesimal calculus in the theory of series, in the calculus of variations, in probability theory, and in mechanics.

BERNOULLI, John (1667 – 1748)

John Bernoulli, brother to Jacob, was born at Basel, and set out to study medicine, but actually studied mathematics with his brother. In 1691 – 92 he was in Paris; in 1695 he accepted the Chair of Mathematics at the University of Groningen in the Netherlands. In 1705 he succeeded his brother at the University of Basel. Less original than his brother, he was, like him, one of the most active propagandists of the infinitesimal calculus and its applications.

BERNSTEIN, Felix (1878 – 1956)

Bernstein was born at Halle, and was a pupil of Cantor in that town, and then of Hilbert and Klein at Göttingen, where he became professor of statistical mathematics in 1911, until the Nazi régime deprived him of his chair in 1934. He then taught in several American universities. His work is on set theory and mathematical genetics.

BEZOUT, Etienne (1739–1783)

Bezout was born at Nemours. From 1758 he was a member of the Académie des Sciences. In 1763 he became teacher and examiner in mathematics of the future officers of the Gardes du Pavillon et de la Marine and of the Corps d'Artillerie. His writings concern algebraic equations and their applications to the early stages of algebraic geometry.

BOLYAI, János (1802–1860)

Born at Koloszvár, in Hungary, the son of a mathematics teacher and former fellow-student of Gauss, Bolyai chose a military career. In 1833, on being discharged as a semi-invalid, he first returned to his father's house, then took up residence on an estate at Domáld. Finding that he had been anticipated by Gauss in his discovery of non-Euclidean geometry, he published nothing further in mathematics.

BOLZANO, Bernard (1781–1848)

Bolzano was born in Prague, and studied theology and mathematics at the University of his native town. He was ordained priest in 1804, and in 1805 was appointed to the chair of philosophy and of religion, which had just been created by the Emperor of Austria. He was dismissed from his post in 1819 because of his non-conformist ideas, and his works were put on the Index. His original mind, very much in advance of his age, conceived many ideas which were developed subsequently, but he could not himself extract all their implications for lack of technical means.

BOMBELLI, Rafael (1526–1572)

Bombelli was born at Bologna, and was an engineer by profession. His treatise on algebra, much clearer and more systematic than that of Cardan, influenced the succeeding generation by its notation and by its treatment of "imaginary" numbers.

BOOLE, George (1815–1864)

Born in Lincoln (England), Boole was self-taught in mathematics. He began to teach from the age of sixteen years, and in 1835 created his own school in Lincoln. From 1849 he was professor at the new Queen's College in Cork

(Ireland). He was elected to membership of the Royal Society in 1857. He must be considered the founder of mathematical logic.

BOREL, Emile (1871 – 1956)

Borel was born at Saint-Affrique in the Aveyron and was a pupil at the Ecole Normale Supérieure. He taught at the University of Lille, at the Ecole Normale Supérieure, and then at the Sorbonne from 1909. He was a member of the Chamber of Deputies from 1924 – 36, and, in 1925, was the Minister of the Navy. He founded the Centre National de la Recherche Scientifique, and contributed to the planning of the Institut Henri Poincaré, acting as its director from 1928 until his death. He was a member of the Académie des Sciences from 1921. He was one of the first to apply Cantor's ideas to analysis and to probability theory.

BRAHMAGUPTA (598 to after 665)

Brahmagupta probably lived in Rajasthan (India). He left several works on astronomy and mathematics.

BROUWER, Luitzen Egbertus Jan (1881 – 1966)

Brouwer was born at Overschie, and spent his whole working life at the University of Amsterdam, where he gained his doctorate in 1907 and taught from 1909 – 51. His writings on algebraic topology of 1910 – 12 made him famous, but the rest of his career was dedicated to the development of "intuitionist" mathematics.

BUNIAKOWSKI, Victor Jakovlevich (1804 – 1889)

Buniakowski obtained a doctorate at Paris in 1825, and taught at the University of St. Petersburg from 1846. He was a member of that city's Academy of Science from 1828.

CAMPANUS (first quarter of the thirteenth century to 1296)

Probably born at Novara (Italy); editor of and commentator on works of astronomy and on Euclid's *Elements* .

CANTOR, Georg (1845 – 1918)

Cantor was born at St. Petersburg to German parents, and was a university student first at Zurich and then at Berlin, where Weierstrass was his teacher. From 1869 Cantor taught at the University of Halle. In 1890 he founded the Association of German Mathematicians, and became its first president. In 1897 he organized the first International Congress of Mathematicians, at Zürich. After 1884 he suffered sporadically from deep depression, and he died at the psychiatric clinic of the University of Halle. His work on set theory and on topology makes him one of those whose ideas have had the greatest influence on the mathematics of our day.

CARDANO, Girolamo (1502 – 1576)

Cardano was born at Pavia, and studied medicine at Pavia and at Padua. From 1534 he taught mathematics in a Milanese school, while practising medicine at the same time. In 1543 he accepted the Chair of Medicine at the University of Pavia, then, in 1562, that at the University of Bologna; but, in 1570, Cardano was abruptly found guilty of heresy, arrested and deprived of his post. His mind was encyclopaedic, and his contributions to the theory of algebraic equations brought general ideas into the subject for the first time.

CARTAN, Elie (1869 – 1951)

Cartan was born at Dolomieu in the French Alps. He received a government grant to enable him to study at the lycée in Lyons, then at the Ecole Normale Supérieure. He taught at the universities of Montpellier, Lyons, Nancy and Paris. He was elected to membership of the Académie des Sciences in 1931. He is the mathematician who has contributed the most, by his extremely original ideas and by his technique, to what is now called analysis on differentiable manifolds, with its multiple applications to Lie group theory, to global differential geometry and differential topology, to partial differential equations, and to mechanics.

CAUCHY, Augustin-Louis (1789 – 1857)

Cauchy was born in Paris, and, after attending the Ecole Polytechnique, passed through the Ecole des Ponts et Chaussées, and contributed as an engineer to various public works. In 1813 he returned to Paris, and taught from 1815 at the Ecole Polytechnique, and later at the Faculté des Sciences and at the Collège de France. In 1816 he was appointed member of the Académie des Sciences. After the revolution of 1830 Cauchy went into exile in Turin, then in Prague, where he became tutor to the grandson of Charles X. In 1838 he returned to France, resumed his work at the Académie, and in 1848 resumed his chair at the Sorbonne. The most prolific of mathematicians after Euler, he published like him in all areas of pure and applied mathematics. He was the law-giver of analysis during the first half of the nineteenth century, and was responsible for most of its progress. He was also the founder of elasticity theory.

CHEVALLEY, Claude (1909–1984)

Chevalley was born in Johannesburg (Transvaal), where his father was Consul General of France. He was a student at the Ecole Normale Supérieure from 1926–1929. He taught at the Universities of Strasbourg and Rennes from 1936–1938, then in the United States, at the University of Princeton (1940–48) and at the University of Columbia in New York (1948–55). On returning to France, he was professor at the Sorbonne from 1955 until his retirement in 1978. He was a correspondent of the Académie des Sciences. To him are owing important contributions to the theory of algebraic numbers and in particular to the theory of algebraic groups.

CHUQUET, Nicolas (second half of the fifteenth century)

Chuquet was the author of *Triparty in the Science of Numbers* , which was remarkable for the novelty of its system of notation. Almost nothing is known of his life.

CLAIRAUT, Alexis-Claude (1713–1765)

Clairaut was born in Paris, the son of a teacher of mathematics. He was remarkably precocious, and was elected member of the Académie des Sciences at the age of eighteen. He inaugurated the study of skew curves, and contributed to the development of the infinitesimal calculus in the eighteenth century and to its applications to celestial mechanics.

COTES, Roger (1682–1716)

Cotes was born at Burbage, studied at Cambridge University, and was elected member of the Royal Society in 1711. He collaborated with Newton on the second edition of the *Principia* , and wrote an interesting essay on the integration of "elementary" functions, in which there appears for the first time the formula expressing the logarithm of $\cos \varphi + i \sin \varphi$.

DEDEKIND, Julius Wilhelm Richard (1831–1916)

Dedekind was born in Brunswick, and studied at the University of Göttingen. In 1854–55 he began to teach as *Privatdozent* at Göttingen, where he worked closely with Dirichlet and Riemann. In 1858, Dedekind was asked to go to the Polytechnikum in Zurich, and in 1862 he became Professor at the Polytechnikum in Brunswick, where he ended his career. Along with Kummer and Kronecker, he was the creator of the theory of algebraic numbers, and, in collaboration with H. Weber, he provided the first completely algebraic treatment of algebraic curves. Along with Cantor and Hilbert, he is the originator of modern concepts in mathematics.

DEL FERRO, Scipione (1465 – 1526)

Del Ferro was born and died at Bologna, where he was a lecturer at the University from 1496 to 1526. He left no work, either in print or in manuscript.

DESCARTES, René du Perron (1596 – 1650)

Descartes was born at La Haye, in Touraine. He graduated in law from the University of Poitiers, and studied mathematics in Paris under the direction of Mydorge and Mersenne. In 1617 he joined the army of the Prince of Orange, and for nine years alternated between serving in various armies and spells of dissipation in Paris. In 1628 he settled in Holland, and in 1649 he accepted an invitation from Queen Christina of Sweden. He died of pneumonia shortly after his arrival in Sweden. With his comprehensive intellect, he was an innovator in all areas: independently of Fermat, he created the method of coordinates in mathematics, after introducing the system of algebraic notation used today.

DIOPHANTUS

Practically nothing is known of the life of Diophantus. He probably lived in Alexandria, about the middle of the third century A.D.

DIRICHLET, Gustav Peter Lejeune (1805 – 1859)

Dirichlet was born at Düren in Germany, and studied in Paris from 1822 to 1826, earning his living by teaching in the family of General Foy. From 1826 – 1828 he taught at the University of Breslau, then from 1829 to 1855 at the University of Berlin. In 1855 he succeeded Gauss at the University of Göttingen. In 1831 he was elected a member of the Berlin Academy of Sciences. He produced important work in analysis and in the theory of algebraic numbers, and he created the analytic theory of numbers.

ERATOSTHENES (ca. 274 – ca. 195 B.C.)

Eratosthenes was born in Cyrene, and spent most of his career in Alexandria, where he was chief librarian at the famous Museum.

EUCLID (about 295 B.C.)

Very little is known about Euclid's life. He probably studied at Athens, and he founded the Alexandrian school of mathematics.

EUDOXUS (ca. 400 – ca. 347 B.C.)

Eudoxus, who was born at Cnidus, studied with Archytas of Tarentum and with Plato, and travelled to Egypt; he taught at Cyzicus, and ended his career in his native town. One of the most remarkable innovators of the ancient world, he was almost certainly the inventor of the theory of ratios and of the method of exhaustion, and he is credited with the first model of the movements of the planets, created with the help of rotating spheres.

EULER, Leonhard (1707–1783)

Euler was born at Basel, where he obtained his doctorate of philosophy in 1723. In 1726 he accepted an invitation from the St. Petersburg Academy of Sciences, and he set out in 1727 to become professor of physics, then of mathematics there. In 1738 he lost the use of his right eye. In 1741, Euler left St. Petersburg for Berlin, where, for twenty-five years, he was director of the mathematics class at the Academy. A conflict with the King of Prussia caused Euler to return, in 1766, to St. Petersburg. A little while later he became blind. His work is the most voluminous in the history of science, and it touches on all scientific and technical subjects. Along with Lagrange, Euler dominated eighteenth-century mathematics through the variety and richness of his discoveries.

FAGNANO del Toschi, Giulio Carlo (1682–1766)

Fagnano was born at Sinigaglia in Italy into an aristocratic family, and was nominated *gonfaloniere* of Sinigaglia in 1723. He was self-taught in mathematics. The results which he obtained on the "lemniscate" arcs attracted the attention of Euler, who generalized them and thus discovered the first results of the theory of elliptic functions.

FERMAT, Pierre de (1601–1665)

Fermat was born at Beaumont, the son of a leather-merchant who was rich enough to allow Pierre to study law at the University of Toulouse. After receiving his bachelor's degree in 1631, at Orléans, Fermat bought the office of *conseiller* in the Toulouse *parlement* . From 1648 he was a member of the Chambre de l'Edit at Castres. Undoubtedly the profoundest mathematician of the seventeenth century, Fermat originated, with Pascal, the theory of probability, and discovered, before Descartes, the method of coordinates. He was the first to provide a general method for the determination of the tangents to a plane curve; but it was above all in number theory that his genius was manifested.

FERRARI, Ludovico (1522–1565)

Born at Bologna, Ferrari became the pupil and collaborator of Cardano. He taught at Milan and at Bologna. He discovered the solution by radicals of a fourth-degree equation by reducing it to a cubic equation.

FOURIER, Jean-Baptiste Joseph (1768–1830)

Orphaned by the age of nine, Fourier became a teacher at Auxerre, his native town, in 1789. After being arrested in 1794, then released after the execution of Robespierre, he returned to Paris to enrol at the Ecole Normale, which was founded and closed in the same year. In 1795 he became an assistant at the Ecole Polytechnique, and in 1798 he followed Monge in Bonaparte's Egyptian campaign. On his return in 1801, Napoleon appointed him prefect of Isère. After the Hundred Days, thanks to a friend, Fourier was nominated director of the Bureau of Statistics of the Seine. In 1817 he became a member of the Académie des Sciences, and, in 1822, its permanent secretary. He was elected to the Académie Française in 1827. His major work was the creation of heat theory, which led him to the theory of what are now known as Fourier series and integrals.

FRAENKEL, Adolf Abraham (1891–1965)

Fraenkel was born in Munich, and studied at the Universities of Munich, Marburg, Berlin and Breslau. From 1916 to 1925 he taught at the University of Marburg, from 1928 to 1929 at the University of Kiel, and from 1929 to 1959 at the Hebrew University of Jerusalem. His work has to do with logic and set theory.

FRECHET, Maurice (1878–1973)

Fréchet was born at Maligny, and was a student at the Ecole Normale Supérieure. He taught successively at the Universities of Poitiers (1910–19), Strasbourg (1920–27) and Paris (1928–40). He was a member of the Académie des Sciences. He introduced many new ideas in functional analysis.

FREGE, Friedrich Ludwig Gottlob (1848–1925)

Frege was born in Wismar (Germany), and studied at the Universities of Jena and Göttingen, where he obtained his doctorate in philosophy in 1873. From 1879 to 1917, he was professor in the philosophy faculty at Iena. His work has to do with mathematical logic and its applications.

GALILEI, Galileo (1564–1642)

Born at Pisa, Galileo was a student of medicine at the University of that town, but he taught himself mathematics outside the University. He was professor at Pisa from 1589 to 1592, at Padua from 1592 to 1610, and afterwards he lived in Florence and its neighbourhood, in a supervised residence after his trial of 1633. He is the father of modern physics, which he studied by means of experiments guided by mathematical models.

GALOIS, Evariste (1811 – 1832)

Galois was born at Bourg-la-Reine, and was a student at the Collège Louis-le-Grand in Paris. He intended to go to the Ecole Polytechnique, but failed the entrance examination, and was finally accepted at the Ecole Normale Supérieure. As early as 1830, he was excluded from the school because of his republican ideas. He then threw himself into the political turmoil, and was actually put in prison. Galois died from the after-effects of a duel the exact circumstances of which are not known.

GAUSS, Carl Friedrich (1777 – 1855)

Gauss was born in Brunswick in Germany into a poor family, and received, in 1792, a grant from the Duke of Brunswick which enabled him to study at the Collegium Carolinum of Brunswick and at the University of Göttingen (1795 – 1798). In 1799 he obtained a doctorate at the University of Helmstedt, and in 1807 he accepted the post of director of the Observatory at Göttingen, where he stayed until the end of his life. From 1818 to 1825 he directed the triangulation works of Hanover. As omnicompetent as Euler and gifted with even more penetration, he brought new life to all branches of mathematics, unfortunately without publishing all his discoveries.

GIRARD, Albert (1595 – 1632)

Girard was born at Saint-Mihiel, then in the Duchy of Lorraine, but, no doubt as a member of the Reformed Church, he had to settle in the Netherlands. He probably studied at the University of Leiden, and was an engineer in the army of Frederick Henry of Nassau, Prince of Orange.

GÖDEL, Kurt (1906 – 1978)

Gödel was born at Brünn in the Austro-Hungarian Empire (now Brno in Czechoslovakia), and studied at the University of Vienna. His thesis was accepted in 1929, and he received his *Habilitation* in 1933. In 1939 he left to work at the Institute of Advanced Study at Princeton, where he became Professor in 1953.

GRASSMANN, Hermann (1809 – 1877)

Born at Stettin, Grassmann began studying theology at Berlin, before teaching mathematics at Berlin and, from 1842, at Stettin, in the secondary sector. His ideas about linear algebra and exterior algebra were not understood during his lifetime, and at the end of his life he devoted himself to linguistics, and became a highly regarded specialist in Sanscrit.

HADAMARD, Jacques (1865 – 1963)

Hadamard was born at Versailles, and after studying at the Ecole Normale Supérieure, taught at the Lycée Buffon in Paris (1890 – 93), at the University of Bordeaux (1893 – 97), at the Sorbonne (1897 – 1909), at the Collège de France (1909 – 37), at the Ecole Polytechnique (1912 – 37) and at the Ecole Centrale (1920 – 37). From 1912, Hadamard was a member of the Académie des Sciences. His work is mainly devoted to analysis and its applications.

HAMILTON, William Rowan (1805 – 1865)

Hamilton was born in Dublin, and entered Trinity College, Dublin, in 1823. In 1827, without having obtained any degree, he was appointed Astronomer Royal of the Dunsirk Observatory. He became a member of the Irish Royal Academy in 1832, and was its President from 1837 to 1845. We are indebted to him for important work in mechanics and optics.

HARDY, Godfrey Harold (1877 – 1947)

Hardy was born at Cranleigh (in England), and was admitted to Trinity College, Cambridge, in 1896. He studied and taught there until 1919, at which date he was appointed professor at Oxford. He spent the year 1928 – 29 at Princeton, and returned to Cambridge in 1931, as professor. He retained this chair until his retirement in 1942. Partly in collaboration with J.E. Littlewood and S. Ramanujan, he published important work in analysis and in the analytic theory of numbers.

HENSEL, Kurt (1861 – 1941)

Hensel was born at Königsberg, and studied at Bonn and Berlin, where he took his doctorate in 1884. He taught at first in Berlin, then, from 1901, at the University of Marburg. He introduced original ideas into number theory, the importance of which was not recognized until after 1920.

HERMITE, Charles (1822 – 1901)

Hermite was born at Dieuze (Lorraine), and was admitted in 1842 to the Ecole Polytechnique, but was turned down a year later on account of a congenital defect in his right foot. Hermite then set himself to follow a teaching career. From 1848 to 1876, he taught at the Ecole Polytechnique, and from 1869 to 1897 at the Paris Faculty of Science. He was a member of the Académie des Sciences from 1856. His work was equally outstanding in algebra, number theory and analysis.

HILBERT, David (1862–1943)

Hilbert was born at Königsberg, and studied from 1880 to 1884 at the University of his native town (with the exception of his second semester, which he spent at Heidelberg). He journeyed to Leipzig and to Paris, and in 1886 he qualified as a *Privatdozent* at the University of Königsberg. In 1895 he gained a chair at Göttingen, which he retained until his retirement. An omnicompetent mathematician, he is, with Poincaré, the figure who has had most influence on twentieth-century mathematics.

HIPPIAS of Elis (about 400 B.C.)

Only the titles of Hippias's works are known, and not much of his life. He seems to have been a "sophist", of encyclopaedic knowledge. Some commentators attribute to him the invention of the transcendent curve known as the "quadratrix", which can be used for the geometric construction of the trisection of an angle and of the area of a circle.

HIPPOCRATES of Chio (second half of the fifth century B.C.)

What is known of Hippocrates is limited to a few anecdotes. He is credited with having written one of the first treatises on geometry.

HÖLDER, Otto Ludwig (1859–1937)

Hölder was born at Stuttgart, studied at Berlin with Weierstrass, and obtained his doctorate in 1882 at Tübingen. He was *Privatdozent* at Göttingen, then professor at Königsberg (1894–99) and at Leipzig from 1899. He left interesting works in analysis and in group theory.

JACOBI, Carl Gustav Jacob (1804–1851)

Jacobi was born at Potsdam (in Germany), the son of a Jewish banker. After his thesis had been accepted at the University of Berlin, he became a convert to Christianity so that he could qualify for the *Habilitation* . In 1826, he was invited to the University of Königsberg. In 1843 the state of his health obliged him to travel to Italy. On his return in 1844, he took up a chair in Berlin. Jacobi was a member of Berlin Academy of Sciences. Known best for his discovery of elliptic and hyper-elliptic functions, he also published many articles on analysis, the calculus of variations and number theory.

JEVONS, William Stanley (1835–1882)

Jevons was born in Liverpool, and studied at University College, London. He taught at Manchester and Liverpool, then at University College from 1876.

His works are concerned with logic and political economy. He was a member of the Royal Society from 1872.

JORDAN, Camille (1838 – 1921)

Jordan was born at Lyons, studied at the Ecole Polytechnique, and followed the profession of an engineer until 1885. From 1873 until his retirement in 1912, he taught at the Ecole Polytechnique and at the Collège de France. He was elected a member of the Académie des Sciences in 1883. His work on group theory dominated this branch of mathematics for all the concluding period of the nineteenth century, and his *Cours d'analyse* remained a model for several generations.

KEPLER, Johannes (1571 – 1630)

Kepler was born at Weil-der-Stadt (Würtemberg). He studied at the University of Tübingen, and then taught mathematics at Graz (in Austria). An edict soon drove him away from there on account of his Protestant religion, and in 1601 he succeeded Tycho Brahe at Prague as astronomer to Rudolf II. The Emperor Mathias (1612 – 19) appointed him as mathematician of Upper Austria with a residence at Linz, a post which he retained under Ferdinand II. In spite of his lofty official patronage, Kepler's financial situation was insecure, and he died at Regensburg worn out by over-work and misery.

KILLING, Wilhelm Karl Joseph (1847 – 1923)

Killing was born in Burback (Germany) and received his university education at Münster and at Berlin. He was a teacher at the high school of Brilon (1878 – 1882), at the high school of Braunsberg (1882 – 1892) and at the University of Münster (1892 – 1920). His main work was the determination of all simple complex Lie groups.

KLEIN, Felix (1849 – 1925)

Klein was born in Düsseldorf and studied at the Universities of Bonn, Göttingen, Berlin and Paris. He taught at the University of Erlangen (1872 – 75), at the Munich Polytechnikum (1875 – 80), at the University of Leipzig (1880 – 86) and finally at Göttingen (1886 – 1913). In 1895 he founded the great *Enzyclopädie der mathematische Wissenschaften* .

KOWALEWSKA, Sofia Vassilievna (1850 – 1891)

Born into an aristocratic circle in Russia, Vassielievna married in 1868, so as to be able to study at a foreign university. She studied mathematics at

Heidelberg, then, from 1871 to 1874, followed special courses under Weierstrass and obtained a doctorate "in absentia" from the University of Göttingen. She was, however, unable to obtain a post until 1884, when Mittag-Leffler obtained a chair for her at the University of Stockholm. Her work has to do with analysis and its applications.

KRONECKER, Leopold (1823 – 1891)

Kronecker was born at Liegnitz (Germany), studied at Berlin, Bonn and Breslau and acquired his doctorate in 1845 at the University of Berlin. From 1845 to 1855, he managed some family affairs, then returned, financially independent, to Berlin. In 1861 he was appointed a member of the Berlin Academy of Sciences, and was able to give courses at the University. In 1883 he succeeded Kummer in the chair of mathematics. His papers on algebra, number theory and analysis are of exceptional depth.

KUMMER, Ernst Eduard (1810 – 1893)

Kummer was born at Sorau (Germany), studied at the University of Halle (1818 – 1831), then taught, from 1832 to 1842, at the high school of Liegnitz, where Kronecker was his pupil. He was professor at the University of Breslau from 1842 to 1855, then in 1855 succeeded Dirichlet at Berlin, and, in 1861, founded, with Weierstrass, the first German Seminary of Pure Mathematics. He was a member of the Berlin Academy from 1855. His main achievement was the theory of "ideal numbers" ("valuations" in current language), which made possible the study of algebraic numbers of any degree.

LAGRANGE, Joseph Louis (1736 – 1813)

Lagrange was born at Turin, and in 1755 he was appointed professor at the Royal Artillery School there. Together with friends, he founded a scientific society at Turin. In 1766, he accepted the directorship of the mathematical section of the Berlin Academy. In 1787, Lagrange left Berlin for Paris, where he received a salary from the Académie des Sciences. He was a member of the commission for weights and measures and the Bureau des Longitudes from its inception in 1795. He taught mathematics at the Ecole Normale de l'An III, and at the Ecole Polytechnique (1794 – 99). He made important discoveries in all branches of mathematics.

LAMBERT, Johann Heinrich (1728 – 1777)

Born at Mulhouse, Lambert had to leave school at the age of twelve, and taught himself. After pursuing various careers, he was, from 1748 to 1758, tutor in an aristocratic family of Coire (Switzerland). During his stay there he became a member of the Swiss Scientific Society of Basel. At the beginning of the 1760s he had the task of organizing the new Bavarian Academy of Sciences, but in 1764 he left Munich for Berlin. In 1765 he was appointed a member of the Berlin Academy. Endowed with an encyclopaedic mind, he left works on philosophy, logic, mathematics and physics.

LAPLACE, Pierre Simon, Marquis de (1749–1827)

Born a farmer's son at Beaumont-en-Auge, Laplace entered on his career in mathematics at the Ecole Militaire of this small town. Here he began to teach. In 1783 he became examiner of the artillery corps, and was elected, in 1785, to the Académie des Sciences. During the Revolution, he took part in the organization of the Ecole Normale and of the Ecole Polytechnique, and was a member of the Institute from its creation. Bonaparte entrusted him with the ministry of the interior, but only for six months. His most important work concerns probability theory and celestial mechanics.

LA VALLEE POUSSIN, Charles Jean Gustave Nicolas de (1866–1962)

La Vallée Poussin was born and died at Louvain, studied at the University there, and stayed there throughout his entire career as a professor from 1891. His work has to do with analysis and its applications, notably the analytic theory of numbers.

LEBESGUE, Henri Léon (1875–1941)

Lebesgue was born at Beauvais, studied at the Ecole Normale Supérieure (1894–97), then taught at the lycée of Nancy, at the Universities of Rennes (1902–1906) and of Poitiers (1906–1910), and at the Sorbonne. In 1921 he was appointed professor at the Collège de France, and, the following year, was elected a member of the Académie des Sciences. He introduced important new ideas in integration theory, potential theory and algebraic topology.

LEGENDRE, Adrien-Marie (1752–1833)

Legendre was born in Paris, and after his studies at the Collège Mazarin devoted himself to scientific research. He taught mathematics at the Ecole Militaire in Paris (1775–1780), and in 1783 was elected to the Académie des Sciences. During the year 1794 he headed the Commission of Public Instruction. From 1799 to 1813, he was examiner at the Ecole Polytechnique, and, in 1813, he replaced Lagrange at the Bureau des Longitudes. Many important theorems in analysis and number theory are due to him.

LEIBNIZ, Gottfried Wilhelm (1646 – 1716)

Leibniz studied at the University of Leipzig, his native town, and at the University of Jena. From 1667 he was counsellor to the Supreme Court of the Electorate of Mainz, and visited Paris and London. In 1676, he took on the position of librarian at Hanover, which was granted to him by the Duke of Brunswick-Luneburg. In 1700, with the help of the Elector of Brandeburg, he founded a Scientific Society in Berlin. After being appointed private adviser to Peter the Great of Russia, he lived for two years in Vienna (1712 – 14), then returned to Hanover, where he died in total solitude. As omnicompetent as Descartes, his works unfortunately remained unpublished for a long time, notably those in logic and in algebra. The algorithms and notation which he introduced in the infinitesimal calculus had far more influence than those of Newton, and are still in use today.

LIE, Marius Sophus (1842 – 1899)

Lie was born at Nordfjordeide (Norway), and studied at the University of Christiania. He gave private lessons to earn his living, spent the winter of 1869 – 70 in Berlin with Klein, and the summer of 1870 in Paris. In 1872 a chair in mathematics was created for him at Christiania, and in 1886 he succeeded Klein at Leipzig. In 1898 he returned to the University of Christiania. He created out of nothing the theory of the groups which bear his name.

LIOUVILLE, Joseph (1809 – 1882)

Liouville was born at St-Omer into a Lorraine family. A former student of the Ecole Polytechnique and of the Ecole des Ponts et Chaussées, Liouville taught at the Ecole Polytechnique (1831 – 51), at the Collège de France (1851 – 79) and at the Science Faculty of Paris (1857 – 74). He was a member of the Académie des Sciences from 1839 and of the Bureau des Longitudes from 1840. His work has a bearing on many questions in analysis.

LITTLEWOOD, John Edensor (1884 – 1977)

Littlewood was born at Rochester in Kent, and studied at Trinity College, Cambridge from 1903. There he stayed for the whole of his university career (apart from three years as professor at Manchester University), and was professor at Cambridge from 1928 until his retirement in 1950. He produced profound work in analysis and in the analytic theory of numbers.

LOBACHEVSKI, Nikolaï Ivanovich (1792 – 1856)

Lobachevski was born at Nizhzni-Novgorod (Russia), and studied at the University of Kazan (1807 – 12), where he taught and worked all his life. He was Rector there from 1827 to 1846, and founded an observatory there.

MENAECHMUS (middle of the fourth century B.C.)

Probably a pupil of Eudoxus. Nothing is known of his life.

MERAY, Hugues Charles Robert (1835 – 1911)

Méray was born at Chalon-sur-Sâone, and after graduating from the Ecole Normale Supérieure, taught from 1857 to 1859 at the lycée of Saint-Quentin, then retired for seven years to a little village near to Chalon-sur-Saône. In 1866, he taught at the University of Lyons, and, from 1867, he was professor at the University of Dijon.

MERSENNE, Marin (1588 – 1648)

Mersenne was born at Oizé in Maine, studied at the Jesuit college at La Flèche, and joined the order of Minims in 1611. Apart from some travelling, he lived in Paris in a Minim convent from 1619. Interested in all the sciences, he conducted an immense correspondence with all European men of science, and organized an academy at his convent of which he was the secretary.

MINKOWSKI, Hermann (1864 – 1909)

Born at Alexotas in Russia, Minkowski lived at Königsberg from the age of eight. He received his university education here, apart from three semesters at Berlin. He taught at the Universities of Bonn (1885 – 94) and of Königsberg (1894 – 96), and at the Polytechnikum of Zurich(1896 – 1902). A chair was then created for him at Göttingen. He invented the method in number theory known as "geometry of numbers".

MOIVRE, Abraham de (1667 – 1754)

De Moivre was born at Vitry-le-François, but emigrated to England with his family after the revocation of the Edict of Nantes in 1685. He had, however, received his formative mathematical training in France, at Saumur and Paris. In London, he gave private lessons in mathematics to support himself. He was a member of the Royal Society from 1697. His principal work was in probability theory.

MONGE, Gaspard (1746 – 1818)

Monge was born at Beaune, the son of a merchant, and, in 1765, was admitted to the Ecole Royale du Génie at Mézières as a technician. From 1766 to 1784, he taught mathematics there. In 1780, he was elected a member of the Académie des Sciences, and, in 1783, was appointed examiner of the naval cadets. A supporter of the Revolution, Monge became Minister of the Navy (1792–93), was on the Committee of Public Safety, founded the Ecole Polytechnique and organized the expedition to Egypt for Bonaparte. The Restoration deprived him of all his titles and offices. His important works are concerned with differential geometry and the theory of partial differential equations.

MONTEL, Paul Antoine Aristide (1876–1975)

Montel was born in Nice, and was a student at the Ecole Normale Supérieure, gaining his doctorate in 1907. He taught at the Ecole Polytechnique and at the Paris Faculty of Science (1911–46). Concurrently, he was professor at the Ecole Nationale Supérieure des Beaux-Arts, Director of the Ecole Pratique des Hautes Etudes, President of the Palais de la Découverte and member of the Académie des Sciences from 1937. His work has to do with analysis.

NEWTON, Isaac (1642–1727)

Newton was born at Woolsthorpe (England) after the death of his father. After being trained at Trinity College, Cambridge, he was appointed professor there in 1669, succeeding I. Barrow. In 1696 he left Cambridge to become Director of the Royal Mint in London. In 1699 he became a member of the Council of the Royal Society, and in 1703, its President. He kept this post until the end of his life. Founder of dynamics and of celestial mechanics, he was also the first to invent a notation and general algorithms for the infinitesimal calculus.

NOETHER, Amalie Emmy (1882–1935)

Emmy Noether was born at Erlangen, the daughter of the mathematician Max Noether. She studied mathematics at the Universities of Erlangen and Göttingen as a non-matriculated auditor, since women were not admitted as regular students. In 1904, however, she was allowed to matriculate at Erlangen, and she was able to acquire her doctorate there in 1907. It was only after many interventions in her favour that she was enabled in 1919 to obtain her *Habilitation* . She taught at Göttingen without ever occupying an official position, until 1933, the date of her enforced retirement at the hands of the National Socialist government. She took refuge in the United States, and worked at the college of Bryn Mawr and at the Princeton Institute for Advanced Study. Her influence was a determining factor in the creation of what has been called "modern algebra".

OHM, Martin (1792–1872)

Ohm was born in Erlangen, and began his academic career in 1811 at the University of Erlangen. After teaching at the high school in Thorn (1817–1821), he became lecturer, then professor of mathematics at the University of Berlin. Concurrently, he taught at the Architekturschule (1824–31), at the Vereinigte Artillerie- und Ingenieurschule (1833–54) and at the Allgemeine Kriegschule (from 1826).

ORESME, Nicole (about 1320–1382)

Oresme was probably born near Caen. He studied at the Collège de Navarre at the University of Paris, and taught there until 1362. After that he was a member of the chapter of Rouen Cathedral until 1377, the date at which he became Bishop of Lisieux. His works contain the seeds of mathematical ideas which were not developed until the seventeenth century.

PASCAL, Blaise (1623–1662)

Born at Clermont-Ferrand, Pascal came to Paris in 1631 with his father, and from 1635 attended the Parisian Academy of Mersenne. In 1640 he followed his father to Rouen, where the whole family was converted to the austere religion of Port-Royal. Pascal returned in ill-health to Paris in 1647, and then entered on what is called his "worldly" period, filled with intensive scientific activity, followed by a second "conversion". After 1654, Pascal devoted himself to a militant Christian life, helping the Jansenists in their conflict with the Jesuits. From 1658, he was extremely ill. He had exceptional mathematical gifts, of which he gave proof in many problems.

PASCH, Moritz (1843–1930)

Pasch was born at Breslau (Germany), and studied at the University of his native town, then at Berlin, under the influence of Weierstrass and Kronecker. He spent all his academic career at the University of Giessen (1870–1911), where, after 1873, he taught a rigorously axiomatic form of geometry.

PEANO, Giuseppe (1858–1932)

Peano was born at Spinetta (Italy), and lived in Turin from the age of twelve or thirteen. There he studied and followed his academic career. From 1890 he was professor of calculus, and from 1886 to 1901 he taught concurrently at the Military Academy. He was a member of the Turin Academy of Sciences. He was the head of a school which participated actively in the formation of twentieth-century mathematical concepts.

POINCARE, Jules Henri (1854–1912)

Poincaré was born at Nancy, was admitted in 1873 to the Ecole Polytechnique, and studied engineering at the School of Mining. He pursued this activity briefly while preparing his doctoral thesis. He taught at the University of Caen (1879–1881) and at the University of Paris (1881–1912). He was a member of the Académie des Sciences from 1887, and was elected to the Académie Française in 1908. With a genius equal to that of Gauss, and just as universal, he dominated every part of the mathematics of his time.

PROCLUS (about 410–485)

Proclus was born in Byzantium, and studied at Alexandria, then at Athens, where he was a member of the Platonic Academy, and its head until his death. Though without originality, he left a precious "Commentary" rich in details about Greek mathematics, from Thales until his own time.

RAMANUJAN, Srinivasa Aaiyangar (1887–1920)

Born at Kumbakonam near Madras, Ramanujan's enthusiasm for mathematics made him neglect his other studies, and in order to survive he had to accept a post as a petty functionary in the port of Madras. On his communicating his discoveries to G.H. Hardy, the latter was able to arrange for him to come to England in 1914, and wrote many papers in collaboration with him. Becoming seriously ill, Ramanujan had to return to Madras in 1919. He had been elected member of the Royal Society in 1918. Gifted with an extraordinary talent in number theory, he left notebooks full of formulae which make us wonder how he discovered them, without being able to give real proofs.

RICCATI, Jacopo Francesco (1676–1754)

Riccati was a Venetian noble, who studied law at the University of Padua and there became interested in mathematics. Refusing very brilliant positions, such as that of President of the Academy of Sciences at St. Petersburg, he devoted himself entirely to his studies. He often served as expert adviser to the Venetian senate in the construction of dykes and canals. His work had to do with the nascent theory of differential equations.

RIEMANN, Georg Friedrich Bernhard (1826–1866)

Riemann, born in Breselenz (Germany), was a student at Göttingen and at Berlin. He acquired his doctorate at Göttingen in 1851 and his *Habilitation* in the same place in 1853. There he taught and, in 1859, succeeded Dirichlet in the Chair of Mathematics. Having contracted tuberculosis, he died in the course of a journey to Italy. A visionary of genius, his ideas, even though not accompanied by proofs, were an inspiration to mathematicians for a whole century.

ROBERVAL, Gilles Personne de (1602–1675)

Roberval was born near Senlis in very humble circumstances. On arriving in Paris in 1628, he introduced himself into the circle of scientists around Mersenne, and in 1634 he won the competition for the Ramus Chair at the Collège de France, which he retained all his life. In 1655 he succeeded Gassendi in the Chair of Mathematics. He belonged to the Académie des Sciences from its foundation in 1666. He was one of the initiators of the infinitesimal calculus, before it was codified by Newton and Leibniz.

RUFFINI, Paolo (1765–1822)

Ruffini was born in Valentano (Italy), and studied medicine and mathematics at the University of Modena. On graduating in 1788, he was immediately appointed professor. After the occupation of Modena by Napoleon's troops in 1796, Ruffini refused, in 1798, to swear fealty to the Republic and was relieved of all his official duties, but still continued to practise medicine. After the fall of Napoleon, Ruffini occupied the Chairs of Applied Mathematics and Practical Medicine until his death.

SACCHERI, (Giovanni) Girolamo (1667–1733)

Saccheri was born at San Remo (Italy), and, in 1685, entered the Jesuit order and studied philosophy and theology at the Jesuit College of Brera. In 1694 he was sent to teach philosophy, first in Turin, then in Paris. From 1699, he taught at the University of Pavia, and occupied the Chair of Mathematics there until his death.

SCHERING, Ernst Christian Julius (1833–1897)

Schering was born in Sandbergen in Germany, and lived at Göttingen from 1852, the year of his acceptance into the University. He was professor of mathematics and astronomy there, as well as Director of the Observatory of Terrestial Magnetism, created by Gauss.

SCHREIER, Otto (1901–1929)

Schreier was born in Vienna, obtained a doctorate from the University of that town, and was *Privatdozent* at the University of Hamburg from 1926 until his premature death from blood-poisoning. He had already produced some remarkable work on algebra and group theory.

SCHRÖDER, Friedrich Wilhelm Karl Ernst (1841 – 1902)

Schröder was born in Mannheim (Germany), and studied at the Universities of Heidelberg and Königsberg. After teaching at the Eidgenössische Polytechnikum at Zurich (1865 – 74), at Karslruhe, Pforzheim and Baden, Schröder accepted, in 1874, a post at the Technische Hochschule of Darmstadt and, in 1876, at the Technische Hochschule of Karlsruhe. In 1890, he became its Director. His work has to do with logic and set theory.

SCHWARZ, Hermann Amandus (1843 – 1921)

Schwarz was born in Hermsdorf in Silesia, and studied chemistry and mathematics in Berlin. In 1867 he was appointed assistant professor at Halle, in 1869, professor at the Eidgenössische Polytechnikum of Zurich and, in 1875, professor at the University of Göttingen. In 1891 Schwarz succeeded Weierstrass at the University of Berlin. He was a member of the Prussian and Bavarian Academies of Sciences. He produced important work on analysis.

SIEGEL, Carl Ludwig (1896-1981)

Siegel was born in Berlin, and studied at Berlin and Göttingen. He was professor at the Universities of Frankfurt (1922 – 37) and Göttingen (1938 – 40). In 1940 he benefited from a round of conferences in Denmark and Norway to escape the Nazi régime, and became professor at the Institute for Advanced Study at Princeton from 1940 to 1951. He became professor at Göttingen again in 1951, and retired in 1959. He was one of the greatest specialists in number theory in the twentieth century.

SKOLEM, Albert Thoralf (1887 – 1963)

Skolem was born at Sandsvaer (Norway), and studied at the University of Oslo. After a study trip to the Sudan, he completed his training at Göttingen, then returned to Oslo to teach there (1916 – 30 and 1937 – 50). From 1930 to 1938 he did independent research at the Christian Michelsen Institute in Bergen. His work is concerned with algebra, number theory and logic.

SMITH, Henry John Stanley (1826 – 1883)

Smith, who was born in Dublin, lived in Oxford after 1840. He studied at Balliol College there. He was elected professor of geometry in 1860, and directed the University museum from 1874. In 1861 he became a member of the Royal Society, and, in 1877, Chairman of the Meteorological Council in London. His work has to do with number theory and elliptic functions.

STEVIN, Simon (1548 – 1620)

Stevin was born in Bruges, the illegitimate child of two rich citizens of that town; but little is known of his life until 1583, the date at which he entered his name at the University of Leiden. He was regarded as an "engineer", and advised the Prince of Orange, Maurice of Nassau, on questions relating to the army and navigation. Of encyclopaedic mind, he left numerous works on very diverse subjects.

STIFEL, Michael (1487 – 1567)

Stifel weas born at Esslingen (Germany), was ordained priest in 1511, and was one of Luther's disciples. In 1535 he signed on at the University of Wittenberg, and gave courses in theology and mathematics at the Unversities of Königsberg and Jena. His works on algebra and arithmetic introduce better notation than that of his predecessors, and contain the first seeds of the idea of logarithms.

THALES (about 625 to about 547 B.C.)

Thales lived at Miletus, and was credited in Classical times with having invented the first proofs in geometry.

VIETE, François (1540 – 1603)

Viète was born at Fontenay-le-Comte, and studied law at the University of Poitiers. In 1564 he entered the service of Antoinette d'Aubeterre as her tutor. From 1570 to 1573 he was in Paris, where Charles IX appointed him as adviser to the *parlement* of Brittany at Rennes. In 1580, he became *maître des requêtes* to the *parlement* of Paris and royal privy counsellor. Banished from the court from 1584 to 1589, he was recalled by Henry III, and became counsellor to the *parlement* at Tours. During the war against Spain, Viète deciphered intercepted letters in code for Henry IV. He was discharged in 1602.

VON NEUMANN, Johann (or John) (1903 – 1957)

Von Neumann was born in Budapest (Hungary), the son of a rich banker, and received his mathematical training from private instructors. He taught in Berlin (1927 – 29), in Hamburg (1929 – 30) and at the Institute for Advanced Study from 1933. He took part in numerous scientific projects connected with the war effort, such as the construction of the atomic bomb. In 1954 he became a member of the Atomic Energy Commission. One of the most profound analysts of the century, he was also one of the most skilled in the application of mathematics to the other sciences.

WARING, Edward (1736–1798)

Waring was born in Shrewsbury, and studied mathematics at Magdalene College, Cambridge. From 1760 he was professor of mathematics at Cambridge. He was a member of the Royal Society from 1763. His work has to do with algebra and number theory.

WEIERSTRASS, Karl Theodor Wilhelm (1815–1897)

Weierstrass was born in Ostenfelde (Germany), and entered the University of Bonn in 1834, but left it after eight semesters without having passed the examinations. He passed them in 1841 in Münster, and taught thereafter in several high schools. In 1854 he obtained an honorary doctorate from the University of Königsberg, and, in 1856, was appointed professor at the Industry Institute of Berlin. He was associate professor at the University of Berlin from 1856, and obtained a chair there in 1864. He was a member of the Berlin Academy of Sciences from 1856. After Cauchy and Riemann, he succeeded in putting analysis on an entirely rigorous basis, and he was responsible for a whole series of very fine discoveries.

WESSEL, Caspar (1745–1818)

Wessel was born in Vestby in Norway, studied at the University of Copenhagen, and became a surveyor and cartographer. His work on complex numbers, published in 1798, remained unknown for a century.

WEYL, Hermann (1885–1955)

Weyl was born at Elmshorn (Germany), and studied at the University of Göttingen, apart from a year in Munich. He taught at Göttingen as *Privatdozent* until 1913, the date of his appointment to the University of Zürich. In 1930 he accepted a chair at the University of Göttingen, but refused in 1933 to remain in Nazi Germany, and accepted a post at the Institute for Advanced Study, Princeton. The most gifted of Hilbert's students, he was as omnicompetent as his master. His finest papers relate to Lie groups, but he also contributed numerous original ideas to the analytic theory of numbers, to functional analysis and to differential geometry.

ZARISKI, Oscar (1899 – 1986)

Zariski was born at Kobryn (Russia), studied at the Universities of Kiev, Pisa and Rome, where he was a pupil of Castelnuovo, and obtained a doctorate in 1924. He taught at the the Universities of Johns Hopkins, Baltimore (1927 – 1945), of Illinois (1946 – 47) and Harvard from 1947. He was one of the founders of algebraic geometry on an arbitrary field, using only commutative algebra.

ZERMELO, Ernst Friedrich Ferdinand (1871 – 1953)

Zermelo was born in Berlin, studied mathematics, physics and philosophy at Berlin, Halle and Freiburg, and in 1894 had a thesis accepted at Berlin. He taught at Göttingen (1899 – 1910) and at Zürich (1910 – 16). Obliged to retire for reasons of health, he went to live in the Black Forest, and in 1916, he was appointed honorary professor of the University of Freiburg. Disapproving of Hitler's régime, he left the University in 1935, but rejoined it in 1946. He worked in mechanics and the calculus of variations, but his most important work concerns set theory.

2. Standard Notation

The following standard notation is used throughout the book

N^*: set of natural numbers 1, 2, 3, ...

$N = N^* \cup \{0\}$: set of integers ≥ 0.

Z: set of integers, positive, negative, or zero.

Q: set of rational numbers, positive, negative, or zero; they are written m/n, where $m \in Z$ and $n \in N^*$.

R: set of real numbers (or the real line).

R^2: the plane.

R^3: usual space of three dimensions.

C: set of complex numbers.

$x \in E$: x belongs to E, or is an element of E.

$x \notin E$: negation of $x \in E$.

$A \subset E$: A is a subset of E, or is contained in E.

$A \not\subset E$: negation of $A \subset E$.

(x, y): pair of objects.

$E \times F$: set of pairs (x, y) such that $x \in E$ and $y \in F$.

$E_1 \times E_2 \times \ldots \times E_n$: set of systems $(x_1, x_2, \ldots, x_n)$ such that

$$x_1 \in E_1, x_2 \in E_2, \ldots, x_n \in E_n.$$

E^n: set $E_1 \times E_2 \times \ldots \times E_n$ in the case where all the E_j are equal to E.

$\mathfrak{P}(E)$: set of subsets of E.

$\emptyset$: the empty set.

$f : E \to F, E \xrightarrow{f} F, x \mapsto f(x)$: mapping f of the set E into the set F.

$a \equiv b \pmod{m}$: congruence modulo the integer m, meaning that the integer $a - b$ is an integral multiple of m.

dy/dx, $y'(x)$: value of the derivative of a real function $x \mapsto y(x)$ of a real variable.

$\int_a^b f(x)\, dx$: integral of the real function f over the interval $[a, b]$.

$\mathfrak{S}_n$: symmetric group formed of the permutations of n objects, having order $n! = 1 \cdot 2 \cdot 3 \cdot \ldots \cdot n$. HI(B).tex

3. Index of Terminology

References are given to the relevant Chapter and Section